INVENTORY MANAGEMENT
DEMYSTIFIED

INVENTORY MANAGEMENT DEMYSTIFIED

Anthony Dear

CHAPMAN AND HALL

LONDON • NEW YORK • TOKYO • MELBOURNE • MADRAS

UK   Chapman and Hall, 11 New Fetter Lane, London EC4P 4EE

USA   Van Nostrand Reinhold, 115 5th Avenue, New York NY10003

JAPAN   Chapman and Hall Japan, Thomson Publishing Japan,
              Hirakawacho Nemoto Building, 7F, 1-7-11 Hirakawa-cho,
              Chiyoda-ku, Tokyo 102

AUSTRALIA   Chapman and Hall Australia, Thomas Nelson Australia, 480 La
              Trobe Street, PO Box 4725, Melbourne 3000

INDIA   Chapman and Hall India, R. Sheshadri, 32 Second Main Road,
              CIT East, Madras 600 035

First edition 1990

© 1990 Anthony Dear

Typeset in 11 on 13pt Plantin by
Saxon Printing Ltd., Derby, England
Printed and bound in Great Britain by T. J. Press (Padstow) Ltd,
Padstow, Cornwall

ISBN 0 412 37700 4   0 442 31203 2 (USA)

British Library Cataloguing in Publication Data

Dear, Anthony
Inventory Management Demystified
1. Stock control
I. Title
658. 7'87

ISBN 0–412–37700–4

Library of Congress Cataloging in Publication Data

available

CONTENTS

Rationale   7

1 Systems and Experience in Inventory Control   9

Overview   9
Systems and control   9
Examples of inventory control systems   13
The importance of experience   19
System or experience?   21
In conclusion   24

2 Available Systems of Inventory Control   25

Overview   25
System variety   25
Two-level systems   27
One-level systems   32
Materials requirements planning systems   34
Japanese approaches to control   35
In conclusion   37

3 Mathematics in Inventory Control Systems   39

Overview   39
Systems and mathematics   39
Forecasting   40
Safety stocks   48
Economic order quantity formulation   52
Lead time   53
Sophistication in summary   54
In conclusion   56

4 Designing an Inventory Control System   57

Overview   57
The problem people have in thinking about inventory
    control systems   57
Categorizing the inventory   61

Determining rules and parameters for the ordering categories 67
In conclusion   72

5 Managing the Stock Control System in Operation   73

Overview   73
Fine tuning the system by the use of feedback   73
The role of managers and stock controllers in the
  stock control process   79
Supplier relations   82
In conclusion   86

Concluding Remarks   87

**Appendix 1: Common Algorithms Used for Forecasting
  and Inventory Control**   89
Appendix 2: Calculation of Top-up Level for Monthly Ordering 99
**Appendix 3: Calculation of Safety Stock as a Percentage of
  Total Stock**   100

Index   101

RATIONALE

A fair portion of my life has been spent in warehouses. Over the years, I have worked as a consultant on a wide variety of stock operations and distribution chains – auto parts, paper merchanting, off-licence stocking, mine and manufacturing plant support stores, food distribution, steel merchanting, non-ferrous metal merchanting, fastener distribution, consumer electronics distribution, major airline engineering support, computer spares, book distribution, lube oil distribution, hardware/DIY and others. From my experience, I have come to the conclusion that, in most operations, stock is not controlled as well as it might be. I use the word 'most' here quite deliberately. In simple terms, this means that companies are investing more money in stock than they need to achieve the service they seek to offer their customers. They are also excessively exposed to the risk of accumulating very slow-moving or obsolescent stock. There are several reasons for this. While most operations nowadays possess computerized 'stock control systems', which perform the transaction processing and accounting functions most adequately, the modules concerned specifically with stock control are often deficient. A further factor is that the replenishment activity is supposed to be both 'systematic' and responsive to market trends, a conflict that is difficult to resolve and which in practice often results in the order depending on who is doing the ordering. Educational courses in stock control are often reduced to imperfect presentations of inadequate mathematics, from which many erroneously conclude that salvation is proportional to sophistication. But the main difficulty lies in the manner in which the various levels of management within the organization interface with the stock control process.

Managers, without fail, monitor the monetary value of stocks and react to increases beyond budget or target levels. But many do little more than this – primarily because they are not at all sure what else they should be doing. And here is the crux of the matter. More than anything else, the present state of stock control reflects a failure by management to understand and to put into practice the mechanisms by which ongoing ordering practice is kept in line with company objectives.

In many ways, stock management is where manufacturing management was in the mid 1970s – not realizing its potential. The revolution of the Just-in-Time (JIT) philosophy caused improvements where previously there had been complacency with existing practices and levels of achievement. The six-hour setup of the fender presses of the US automotive company was validated by work study and years of operational experience – until it was learned that Toyota set similar presses in under half an hour.

It is said that JIT is not applicable to warehouse operations with thousands of items, long lead times and unpredictable demand patterns. It is true that it is not realistic in such conditions to expect things to arrive only as the customer wants them. But the underlying philosophy of JIT, characterized by concepts like 'habit of improvement' and 'elimination of wasteful practices' is eminently applicable to most warehouse operations. Most could cope better than they do.

This book is an attempt to assist managers in *thinking through* the meaning of inventory control for their operation and then in working towards the achievement of that control. This is not an easy process because it means thinking about inventory in a manner that is unfamiliar to many managers; there is more detail involved than most like to consider. But a greater understanding of the mechanisms of control is the necessary starting point for any manager with an interest in enhancing that control.

NB: The male pronoun has been used throughout this book for stylistic reasons only. It covers both masculine and feminine genders.

1

SYSTEMS AND EXPERIENCE IN INVENTORY CONTROL

Overview

Stock control is often perceived as one of the more 'systematic' aspects of company operation. The mere fact that we talk so often of the 'stock control system' and that such discussions tend to be peppered with terms like 'exponential smoothing' and 'economic order quantity' reinforces this view. But, in practice, stock control is often better described as being unsystematic rather than systematic. In the first part of this chapter, we will look at a few examples that illustrate this point.

At the heart of any attempt to control stock is the apparent conflict between system and experience. Should the approach be to baseordering decisions on a set of rules that reflect past history or should we rather attempt to seek out and follow developing market trends? We will explore this issue in the second part of this chapter.

Systems and control

The idea of an inventory control system

What is an inventory control system? Perhaps the best place to begin in answering this question is to look at an example of the inventory control function.

Inventory control is concerned with replenishment and this replenishment comes about through ordering. The prime act of inventory control is to place an order on the supplier. But this is not as easy as it might seem.

Suppose you are the inventory control person, responsible for ordering in from an electrical goods wholesaler. How much would you order in the following situation, given that you can place orders every week?

Item number	A45872
Item name	Electric drill
Lead time	2.5 weeks
Item cost	£35.75
On hand	8
On order	9
This order	?
Weekly demand in past 12 months	18 15 12 17 21 4 19 36 15 12 18 29
12-month average	18
6-month average	21.5

I have posed questions like this to numerous audiences of inventory control people and managers. The response is always totally predictable – there is a considerable variety of answers even from members of the same company.

Why is this? When I ask people how they arrived at their answers, they usually describe some calculation or consideration that they used in arriving at the number they wrote down. For example, they might use the average to calculate so many weeks' stock and then subtract the on-hand and on-order balances. Each of them seems to have his own set of rules or guidelines: the answers differ because different people use different rules. So here are three important points:

1. Most people, when asked to derive an order, use a set of rules or guidelines.
2. These rules vary from one person to another, which is why different people without formal guidance in an ordering situation formulate different orders.
3. This variation in orders has nothing to do with knowledge of the market – it arises solely from varying calculations on the same numerical data.

There is a further point here. Which is the right answer? It needs little thought to appreciate that there is no correct answer. Clearly, an order for 500 drills would seem to be excessive by most policies but it would be possible to present arguments in support of orders of 5, 10 and 15 units. A consistent approach that ordered 5 in such circumstances would result in a different overall investment in stock and service to customers than would an approach that ordered 10. But it is no easy matter to ascertain in advance which policy we should use if we wished to achieve an overall service level of 95%. It all seems very vague at this level.

The impact of different ordering patterns

The importance of inventory control or ordering rules is easy to see. The general idea is to keep stocks low and service high. But the actual balance achieved between investment and service is a direct result of the ordering process. Inventory only arrives because it is ordered.

It could be argued that while people might differ slightly in their patterns of ordering, the overall impact of such differences is not significant. Does it really matter in the foregoing example if an order is placed today or in a few days, or if one person orders 10 while another orders 15? This touches on one of the more insidious aspects of the ordering process. Slightly different ordering patterns applied across many items can give rise to quite large variations in the resulting balance struck between inventory investment and service. It is like flying an airplane where a slight error in direction can result in a final position that is miles from the desired destination.

It is not at all easy to ensure that the individual replenishment decisions remain in accord with the overall objectives of the company relating to things like investment in stock. An analogy might help to illustrate this.

Suppose I go shopping and I buy a shirt and some other bits and pieces. I start off with £300.00 in my pocket. Come the end of the day, I only have £10.00. I say to myself 'I couldn't have spent £290.00' – and yet when I count up my purchases I have indeed spent that amount.

It's the same with ordering. It is easy to order what is 'needed' and yet in so doing to spend much more than the company would like us to spend. What seems sensible on an item-by-item basis may add up to a total picture that is undesirable. In particular, a person with a slight bias towards a generous order cover can get a company into a great deal of trouble as time passes – even though if we looked at any one of his individual orders it would not seem unreasonable.

Inventory problems come primarily from slight bias over a period of time rather than from large obvious errors in ordering.

Removing the subjective element

It is not difficult to remove this subjective element from the ordering process. All that is needed is the laying down of a set of rules which are to be followed mechanically. In such a case, of course, there is no need for a stock control person. For instance, if, in the electric drill example, you had been given not only the history and balance data, but also a clearly defined set of rules to follow, then there would not be a variety of possible answers. Suppose the rule was:

> When assets (on hand + on order) fall to one month's stock (based on a 12-month average) order another month's stock.

In this case, following the rule yields only one answer – 18. A group presented with the drill data and the above rule would all arrive at the same ordering decision – provided they followed the rule.

There is nothing new about this. Most company systems – taking customer orders, costing, debtors control, and so on – are conducted in the framework of very tightly defined sets of rules. When we ask the cost accountant to ascertain the cost of a product, we expect him to follow the established company methods for calculating costs – whether he agrees with the company method of allocating overhead or not. Inventory control is in fact unique in that there is so often a lack of clearly defined guidelines.

The essential problem with laying down the rules for inventory control is that no one has sufficient confidence in his ability to settle on the 'right' set of rules; so, no rules are defined and no textbook solutions are sought. As we shall see, these rules are extremely important.

The emphasis in this book

This book is based on some very simple premises that I feel are relevant to most inventory operations:

1. Much of the inventory control process in most companies is unsystematic because of the presence of a high arbitrary or subjective element and/or the absence of clearly defined and sensible rules that are followed. This is true with computer-based systems as much as with manual systems.
2. This is a prime contributing factor to a situation frequently found in organizations – high inventories combined with low service. It also causes excessive obsolescence.
3. To improve the situation, clearly defined rules need to be established and followed sensibly – that is, they should allow for the input of valid market knowledge when it exists.
4. This is no easy task because the 'best' rules cannot be completely established in advance; rather, they must be arrived at by a process of trial and error. Such a process of continual fine-tuning requires feedback systems of some subtlety.

As we proceed, I hope to provide some justification for these opinions and to suggest lines of approach to solutions.

Examples of inventory control systems

Practical systems of control

This section considers some examples of the types of 'systems' people use to control inventory in practice. All are drawn from real-life operations. The pictures drawn are not at all flattering but they are true representations of reality – even though they differ from the concepts that many managers have of their own operations. The purpose of this is, in part, to show that what is often thought of as one of the more 'scientific' aspects of company operation is in fact not so: inventory control in many companies is illogical rather than logical, subjective rather than objective, and subject to erratic input under the guise of market knowledge.

Alpha spares

Alpha Spares is an independent automotive spares operation carrying a range of 25,000 items with a total inventory valuation of £1,000,000.

Inventory control is conducted on a small stand-alone modern computer system which handles order entry, inventory control and accounting functions. There are two ordering people whose prime functions are reordering on the 200 odd suppliers and handling warranty claims.

Inventory control is conducted using a MAX/MIN approach. When assets (on hand + on order − customer backorders) fall below MIN, an order is suggested to bring assets back up to the MAX. The MAX/MIN values are held in the computer and a suggested reorder report produced weekly, by supplier, based on these levels.

The MAX/MIN values are set by the ordering people. They operate under management objectives of turning over the stock four times a year and of providing a 95% service to customers. These objectives guide the manner in which they set the MAX/MIN values.

One of the ordering people described the manner in which he sets MAX/MIN values in the following terms:

> 'We are aiming to turn our stock over four times a year and to provide a 95% service to our customers – that's what we have agreed with management. But our business is different from others – the patterns of demand we experience are very erratic. So we have to

consider how each item is moving when we set the MAX/MIN values. Basically, we set our MAX/MIN values depending on lead time, average movement and the trends in the marketplace.'

Aside from the fact that the computer does not track service in any form – in which case it could be argued that talking of a '95%' service level is rather pointless – this description of how the MAX/MIN values are set is quite simply not supported by the facts of the situation, which are as follows:

- There are no clearly defined rules for setting MAX/MIN values. The objectives – stockturn of 4 and a service of 95% – do not constitute a set of rules.
- Even if there were a set of rules, the physical task of keeping the MAX/ MIN values up to date for all 25,000 items is so time consuming that it could not be done in the time available.(An average of a minute per item would take 417 man-hours – or ten weeks full time – to update them all.) If rules did exist, of course, the sensible approach would be to program them into the computer – not a difficult task, and one that could save much time.
- Because the MAX/MIN values are not up to date, they should not be used as a basis for ordering. Yet the system in place 'suggests' an order when – and only when – the stock on hand and on order falls below the MIN. This logic is clearly heavily dependent on the validity and currency of the MAX/MIN values, and will generally give a rather poor balance between investment and service if they are not kept up to date.

Over the years, I have worked with a considerable number of computer-based inventory operations with large numbers of items where the control parameters (MAX/MIN, order point/order quantity, or whatever) are set manually, as is done in Alpha Spares. Without exception, I have found that these parameters are kept current for, at most, a relatively small number of the fast-moving lines – at most, a few hundred out of several thousand. Many 'inventory control' packages currently on the market opt for this item-by-item manual setting of ordering parameters – possibly because of the difficulty the designers have in arriving at a widely acceptable approach to automatic calculation. Yet from the point of view of *control* – establishing the best achievable balance between investment and service – such approaches are at best loose and often quite inadequate to the task. Ironically, managers with such systems who experience problems in controlling inventory – and most of them do – seem not to appreciate that a large proportion of the difficulty is of their own making.

If we wish to control stock, then we will not do it by choosing an approach that sometimes tells us to order what we do not need and at other times does not tell us to order what we need, simply because we do not keep the ordering parameters up to date.

Beta paper merchants

Beta Paper Merchants are a fine-paper merchanting company. For many years, they have been heavily computerized and their present system has been specifically designed to meet their particular needs.

The computer system incorporates a highly developed inventory control module which was put together by an inhouse operations research team. This is a reorder point/reorder quantity approach. Unlike Alpha Spares, these ordering parameters are kept up to date by being automatically recalculated by the computer at the end of each month. In fact, the calculations used are set out in Fig 1. What do you think of them?

NFCST = 0.8*OFCST + 0.2*LDEM + 0.15(OFCST − LDEM)
NMAD = 0.8*OMAD + 0.25* (LDEM − NFCST)
SAFSTK = 2.2*NMAD + 0.5*NFCST
ROP = LT*NFCST/30 + 1.3*SAFSTK
ROQ = (24*LDEM*OCOST/HCOST)**1/2

where:

NFCST = New forecast (units/month)
OFCST = Old forecast (units/month)
LDEM = Latest month's demand
NMAD = New mean absolute deviation
OMAD = Old mean absolute deviation
SAFSTK = Safety stock in units
ROP = Reorder points in units
ROQ = Reorder quantity in units
OCOST = Ordering cost
HCOST = Holding cost

Figure 1   *Order point and order quantity calculations used by Beta Paper.*

Such a question is not easy for most managers to answer. Often, they have no real opinion because they are not mathematicians or, even if they have a quantitative background, they may not be familiar with inventory control theory. They tend to accept such formulations on faith – to assume that the formulations are at least theoretically sound and even of proven practical worth.

In fact, every single line of the equations set out in Fig. 1 is nonsensical – there is no mathematical logic in any one of them. Moreover, I did not make these up: every one is drawn from a real-life inventory situation. (These are included only to illustrate a point: despite the tendency to get the mathematics wrong, it is unlikely that any one company like Beta Paper would manage to end up with such a complete accumulation of nonsense.)

The point is that mathematical algorithms are often used to calculate inventory control parameters such as reorder point/reorder quantity. The reason for this must be that such algorithms are assumed to be 'better' or more comprehensive than simple approaches – like averaging and weeks of stock – that are understood by most. In fact, there is little evidence that this is so and, worse than that, many practical systems that go down the mathematical route end up, for one reason or another, incorporating nonsense in the guise of sophistication. A key factor here is the tendency to tinker with stock parameter calculations as time passes; while the original algorithms may have been theoretically pure, subsequent changes that try to compensate for perceived deficiencies result in algorithms that have flawed theoretical justification.

Even if they get the mathematics right (and that shouldn't be too difficult as it can be copied directly from the textbooks), the problems posed by a lack of user understanding of the algorithms involved are considerable. In simple terms, the stock controller has the options of accepting a suggestion based on algorithms he does not understand or of overruling it – but how does he know if he should overrule it or not, if he does not understand how the figure was derived. An example will illustrate.

A fastener warehouse used a computer package with a very sophisticated set of algorithms to generate suggested orders. These were then vetted by the order clerk. On investigation, it was discovered that the order clerk ignored the mathematics and used a calculator to work an order to give 'so many weeks' cover based on the history and balances shown on the report. The order clerk did not understand the mathematics and therefore did not trust or use them.

Beta Paper, unlike Alpha Spares, have introduced an automatic update of their ordering parameters. This is very good. But they have opted for a sophisticated mathematical approach that is not well understood by anyone in the company outside the operations research department, and in particular by the people who have to do the ordering. The risks attached to this often outweigh the rather questionable advantages of using it.

Gamma electrical components

Gamma Electrical is, or was, at the forefront of the new technology, and stocks a wide range of electrical and electronic components. Some years ago, the company was aggressively seeking market share in this highly competitive business. Nothing was more frustrating than being out of an item when the customer was there waiting to take it. As the inventory controller was blamed whenever there was an out-of-stock situation, he soon learned to avoid this by making sure he was covered. But while customer satisfaction grew so did the inventory.

Soon there was a directive from the chief executive to reduce stock. At this point in time, because of the over-ordering, there was a relatively high proportion of slow-moving stocks. The position was as indicated in Fig 2(a).

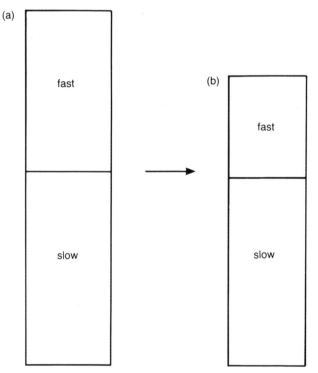

Figure 2   *The effect of a management injunction to reduce stock quickly in a situation which is primarily overstocked in slow and non-moving lines. The reduction is achieved by cutting back on the ordering and this impacts primarily on the fast-moving lines.*

In response to the pressure from the top, the inventory controller cut back on his ordering. But since most of the value of the orders in any time period was for the fast-moving items, this impacted primarily on the items that were selling. So the situation became as depicted in Fig 2(b).

Now, despite the fact that the management objective of a stock reduction has been achieved, a most undesirable state of affairs exists.

- There is a low service being offered in the products that really count – the fast movers that account for a high proportion of the turnover.
- The 'overstock' problem still exists.
- Once a company gets into this situation, it is not easy to get out of it in the short term.

In inventory situations with large numbers of items, this is a very common situation. It is almost inevitable if the rules of inventory control are not clearly defined – or followed – and the inventory controller is subjected to pressures in one period to improve service and in another to reduce stock. Turning the tap on and off like this is extremely disruptive, and almost guarantees a high proportion of slow and non-moving stock coupled with a low service in moving items.

Some conclusions

The foregoing examples give a view of the types of inventory situations that are to be found in practice. They represent the norm rather than the extreme. Many operate without any clearly defined rules of how people are supposed to go about the ordering function. If rules are in place, as in Beta Paper, they may be either inappropriate or incomprehensible to the ordering people.

There are usually objectives that are set by management or reflected in the budget ('Turn the stock over 6 times a year' or 'Keep stock at £1,000,000' or 'Maintain a 95% service level'), but there is no easy way for the ordering person to relate his activity of placing orders to these objectives. For instance, returning to the electric drill example, how many would you order to ensure a stockturn of 6 or to achieve the company's objective of a 95% service level? The answers are anything but obvious. We don't have any clear idea of what investment/service balance is achievable – can we, for example, expect both a stockturn of 6 and a 95% service?

Finally, as shown in the third example, the person doing the ordering is under pressure from both sales (to offer high service) and from general management (to keep stocks low). This makes it difficult for the person concerned to be systematic with his ordering.

The importance of experience

What is experience?

It is important to have a systematic approach to the inventory control function and this mostly involves a clearly defined set of rules to calculate orders on suppliers. But there are situations when such a set of rules should not be followed.

Consider the electric drill example described earlier. On that occasion, we worked out an order based on the balance and history information provided. But suppose we also knew that the prime competitor had just released a new drill which competed directly with our product. It is three-quarters the price and even we are forced to admit that it looks much better than our own product. Moreover, the launch is being conducted with an enormous advertising campaign. Would this not cause us to change our order?

This extra information that is not known to our 'system' but is relevant to ordering is variously known as 'experience', 'market knowledge' or sometimes 'gut feel': this last term is an indication of how tenuous such information often is.

While there is no doubt that market knowledge is essential to the inventory control process, the impact that it has on this process depends largely on the skill of the person who has to interpret it. Ultimately, market knowledge must be expressed as a number – and it is rare that the issue is as clear as was indicated in the electric drill example.

A common argument for experience

One of the more common arguments in favour of 'experience' runs along the following lines:

'Our sales patterns are highly variable and we never know what is going to sell from one month to the next. Moreover, our lead times are highly unpredictable – the same item may take two weeks if our supplier has it in stock and four months if he hasn't. A system can't cope with this situation. We tried once, on computer: it was a dismal failure – far too inflexible – and after a year we threw it out. We have to rely on the experience of our people.'

Now this is an interesting argument. In summary, it says that because we are uncertain of our demand and lead time we cannot use a system but must rely on the experience of our people. Consider an example.

A small steel section has the following sales over the last 12 months:

15   18   511   22   3   233   12   0   632   15   12   56

The lead time can be anything from 3 to 12 weeks and we don't know what it is until we place the order. This is not an easy item to control and a system would not work very well. Just like the man said.

But how does 'experience' or 'market knowledge' help here? It would appear that the prime problem in this case is precisely a lack of knowledge of both demand and lead time. We should not confuse a lack of knowledge with market knowledge – and yet this is frequently done. Because a system is patently unable to cope well with such a situation, it is assumed that an individual using his experience or market knowledge can. But left to their own devices to replenish such an item, many stock controllers would tend to over-order: this is not due to 'experience' but because they do not want to run out of stock. It is precisely in situations like this that a system has something definite to contribute – that is, a middle path that avoids extremes.

In passing, it is clear that if the variation in demand or lead time can be made smaller – by whatever means – then this is an avenue that should be followed up.

The difficulty with market knowledge

While market knowledge is essential to the inventory control process, it is often involves a great deal of guesswork. Here are a few examples of imponderable factors:

- Competitor action.
- An imminent price rise.
- Loss or gain of a major customer.
- Weather change (for agricultural products).
- Local building activity (for steel merchant).

All of these factors could well impact on future patterns of demand and so on the inventory control function. It is not at all easy to gauge their impact in advance. At the same time, they cannot be ignored.

Suppose Beta Paper have a large potential new customer, a major printing works. They do not know yet whether they will land the business or not, but if they do, this customer is likely to want some papers in large quantities at short notice. However, the company will not be able to give this service and still maintain deliveries to other customers unless at least some of the expected papers are ordered in advance. But it is not at all clear what papers or in what quantities the printer is likely to order, if and when he becomes a customer – he has given some broad indications of general usage patterns and types of papers, but few indications about specific sizes and qualities. Even assuming there are good lines of communication between the sales staff who visit the potential client and the inventory control people who order the paper, to reduce such 'iffy' information to specific order quantities is no easy task. As with any attempt to forecast, the results will inevitably be mixed – sometimes right but often wrong.

There is some tendency in inventory control circles to think that because someone 'knows the market', then it follows that this is all that is needed to make this person a stock controller: his 'market knowledge' will guide his actions to a greater or lesser degree of infallibility. This is nonsense.

Consider the case of forecasting. Often inventory control people have to work from a forecast provided by the sales department, and presumably 'market knowledge' is used in deriving the forecast. Yet there is in these cases a common complaint from the inventory control people regarding the 'quality' of such forecasts. This would at the very least seem to indicate that while a high estimation of market knowledge often exists internally within the department that relies on it, this perception is not shared by those that have to use the results of this intuition. If sales people can't estimate the market, why should inventory controllers be any better at it?

System or experience?

The arguments

Should we use a system that has a clearly defined set of rules for deriving ordering parameters and orders or base the ordering on the ordering person's intimate knowledge of the market?

This is a common dilemma. A financial director might argue in the following manner for the benefits of a system:

> 'What we need is a systematic approach to the inventory control function. We need to put the same type of discipline into it that we have in the accounting function. If the inventory control people really know their market so well how is it that we have £250,000 in obsolete stock? We desperately need an objective inventory control system that reflects management objectives in this company.'

Equally persuasively the sales director might put forward a case for the importance of market knowledge.

'Our marketplace is constantly changing. Above all else we need to be able to discern market trends and to follow them quickly. Any "system" is based on history: we need to look to the future, not to the past. The only people who really know what is happening out there are the sales and inventory control people. We must use their knowledge as the prime input to the inventory control process. Only then will we have what the customer wants when he wants it.'

Despite the apparent divergence of views, both the financial director and the sales director want an approach that 'works' – that is, that reflects market trends and yet keeps inventory valuation within budget.

But the question posed earlier – of whether inventory control should be based on systems or experience – is misleading. It is not a matter of choice but of combination: both a system and experience are necessary elements in the inventory control process.

System and experience

How is a systematic approach to be combined with experience so that both can be used in the control of inventory? In simple terms, it is best in most cases to generate a 'suggested order' based on the system and then to overrule it if some valid market knowledge relating to the item or product exists. This approach is illustrated in Fig 3.

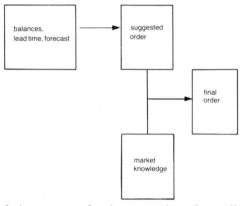

Figure 3   *An ordering strategy that is systematic and yet allows for the input of market knowledge. The system suggests an order which is accepted unless the ordering person is aware of some market factors that lead him to think that the suggested order is inappropriate. This will mostly occur if he expects that the future demand pattern will differ markedly from that reflected in past history.*

In practice

Many stock control people are likely to say:

> 'But that's what we do now. We produce a suggested reorder report and then we go through it and change suggestions that are inappropriate.'

While the practice of printing a 'suggested reorder report' and then having someone go through it is common, almost universal, this is not the same as the procedure outlined in Figure 3. And here is the focus of much that is wrong in inventory control. Often the rules are unclear or the parameters not kept up to date so that 'suggestions' are changed not only in the light of market knowledge but also because they are non-sensical suggestions. Alternatively, the rules by which the 'suggestions' are derived may not be understood by the person who vets the report and so he changes any that 'don't look right': this is not using market knowledge. Moreover, other suggestions are rounded. A '7' becomes a '10', not because the item comes in packs of 10 or the supplier will only deliver this minimum quantity, but simply because of some inner concern for symmetry.

It is also wrong to assume that the overrulings are systematic. In a real-life case, it was possible to ascertain when the stock controller went to lunch merely by looking at the suggested reorder report he had been working on – there was a significant change in the pattern of overrulings half-way through the report. In a different company, there were two stock controllers, one of whom overruled around 50% of the system suggestions while his partner changed only 15%. On investigation, it was found that the prime reason lay not in an enhanced perception of the market possessed by the first controller, but simply in the fact that the second controller liked and trusted the system while the first did not.

The procedure of printing a suggested reorder report and then changing these suggestions only in the light of market knowledge is far from being a widely used approach to inventory control. In most cases, with a simple set of appropriate rules, the proportion of overrulings for reasons of market knowledge should be small – under 10%.

Improving the balance between investment and service

It is not difficult to get people to agree on principles. In the case of inventory control, for example, most people would agree with the idea of using a systematic approach that allows for the input of market knowledge. But this is not what happens in practice.

At the heart of many inventory control problems is the manager's perception of his own operation. He usually thinks it is much more systematic than it is and he has an exaggerated impression of his people's ability to use their knowledge of the market to guide the ordering process. Moreover, the principles and practice of inventory control are often unknown to him. As a result, he is loath to interfere. But merely by concentrating attention on this area and by working away at improving both the system – that is, the rules – and the manner in which market knowledge is incorporated into it, it is possible for a manager to enhance the balance struck between investment and service.

In conclusion

In this chapter, we have looked at some examples of the types of difficulties that beset operating stock control systems. The essential problem in many cases stems from the absence of clearly defined and sensibly followed rules or guidelines for control. If such a condition exists, then it is indicative of a failing on the part of management.

We also saw that there is no conflict between the concept of a systematic rule-based approach to stock control and the need for the ordering decisions to be based on market knowledge. In simple terms, the 'system' provides a history of market knowledge that is used unless the stock controller possesses some specific knowledge regarding a product.

Since the whole process depends on the manager's ability to ensure that a set of formal rules is formulated for his operation, we now need to consider these stock control rules in a little detail. This is done in the following chapter.

2

AVAILABLE SYSTEMS OF INVENTORY CONTROL

Overview

If a manager is to assist his department in the choice of suitable methods for determining order points and so on, then he needs to know the available alternatives. These are simple concepts that can be presented in simple ways, and this chapter attempts to do just that. As we will see, there are really only a small number of options and it is from among these that the manager and his people must choose their 'system'.

System variety

The importance of knowing what is available

For the purposes of our discussion, a stock control system is a set of rules to calculate an order on a supplier. If they are followed, these rules largely determine the ordering pattern and so the actual balance struck between investment in inventory and service to the customer. If we change the rules, then this will impact on both stock investment and service.

How do we go about deriving a set of rules that are appropriate for our operation? Or if we already have some rules in place, how are we to be sure that they are adequate? To begin with, we need to distinguish between two aspects of these rules – the 'equations' and the 'numbers' used in the equations. An order point might be set using the equation:

Order point = Z × average period demand

but for one group of products we might set Z at 4 while for another we might use 3.5. Clearly, before a set of rules are of any use, both the equations and the numbers must be set.

It should be clear that we first of all have to decide what sort of equations we are going to use to control our inventory, and then we need to determine the numbers we are going to use in these formulae. Once we have decided on the equations, we control our stock in response to changing market and company conditions by varying the numbers in the rules; usually, we don't need to rethink the rules.

People are often unclear about the equations and the numbers used in them when they think about stock control. If the 'system' isn't working – that is, people are not happy with the stock/service balance they are achieving – there is some tendency to 'get a new system', instead of the much simpler alternative of refining the equations and the numbers of the existing system.

One way of finding out the available options of stock control systems – which is concerned with the equations relating order points to the other parameters, like lead time and demand on which they depend – is to buy a book on the theory of inventory control. There are a number available. However, for many people, this does not prove a very helpful course of action. Such books tend to be long on mathematics and it is difficult for the lay reader to make much sense of them. In fact, the reading of such books can prove to be a disadvantage because the reader may soon begin to think that inventory control is a matter of mathematics more than anything else. As we shall see, this is not the case: effective inventory control is applied commonsense. But it needs to be based on some knowledge of the available options. These are discussed in the following section.

The two decisions of inventory control

If a manager is asked:

'How many costing systems are there?'

he might not know but he can quickly find out. Any cost accountant would be able to tell him of the options – average cost, LIFO, FIFO, latest cost, and so on – and would be able to explain precisely what each method entails. Costing is highly systematic; that is, the costs are derived according to a very clearly defined set of rules. But ask that same manager:

'How many inventory control systems are there?'

and he will probably have no clear idea nor will he know how to find out. There is a great deal of woolly thinking regarding inventory control generally and inventory control systems in particular. Let's see if we can make some sense out of this.

As a starting point, we note that there are two decisions that need to be made by any inventory control system:

1. When to buy.
2. How much to buy.

The options depend on the way in which these decisions are to be made. The essential choices for stock control systems are:

- Two-level systems, where the user makes both the decisions when to buy and how much to buy.
- One-level systems, where the decision 'when' is imposed on the user: all he has to decide at the 'ordering window' is how much to buy.
- Materials requirements planning (MRP) systems, which are primarily designed to satisfy the materials control needs of manufacturing.

In addition, any adequate discussion of systems of inventory control needs to consider the Japanese approaches that are described by terms such as 'Just-in-Time' and 'Kanban'.

While there is a considerable body of inventory system theory in the operations research literature, comparatively little of this is actually used in real stock control systems. What is found in practice – where rules actually exist – is the repeated use of a small number of standard algorithms that were first used by R. G. Brown in the 1950s and which have survived with little alteration to the present.

In the following sections, each of these three systems is discussed in turn.

Two-level systems

The concept of order point

The most widely used concept of inventory control is that of 'order point', also called 'minimum' or 'reorder level', which is used to determine 'when' to order in a two-level system. The most common two-level systems are:

- Order point/order quantity (OP/OQ) systems.
- MAX/MIN systems.

In the first type of system, once the order point and order quantity are defined, the replenishment activity is conducted as follows:

When assets (on hand + on order − customer backorders) reach the order point, place an order for the order quantity, or sufficient to top up to the order point, whichever is the greater.

The MAX/MIN system is very similar to the order point/order quantity system in that the MIN is the same as the order point; but instead of ordering the order quantity, the order comprises whatever is needed to bring assets up to the MAX. An order point of 60 and an order quantity of 40 is similar to a MIN of 60 and a MAX of 100. The difference between the approaches is illustrated by an asset situation of 55: in the OP/OQ system, 40 would be ordered while in the MAX/MIN system, the order would be for 45.

This is all very simple stuff. The difficulty comes when we begin to think about how we should set the order points and order quantities (or the MAX and MIN values).

Determining the order point

How is the order point or MIN to be set? The order point can depend on a multitude of factors, of which lead time and expected demand are perhaps the most obvious. But what about unit cost, the relative importance of the item, shelf life, expected variability of demand, supplier reliability, and other such factors? Surely it is possible to argue that each of these should be taken into consideration in the determination of the order point?

The idea of an equation for the order point is to establish the quantitative relationship between it and the other quantities that it depends on; but this is no easy task when there are so many factors.

To get some feeling for the task, consider the example shown in Fig 4. This gives the data for a spare part for a commercial vehicle – a VOR (vehicle-off-road item) – which signifies its importance. Other relevant information is also shown.

What would you set as the order point for this item? How would you reflect factors like unit cost and importance, for example, in your formulation? The best way to appreciate the difficulties involved in this task is to do it yourself – don't read on until you have worked out and written down a value for the order point.

This is no easy task for one item in isolation. As this is one of several thousand, each of which requires the setting of an order point, the overall task is quite daunting – especially if we wish to take account of the individual factors that relate to each product.

So how should the order point be set? In fact, this is very like the problem faced by the cost-accounting function in attempting to establish product cost. And the solution is much the same – an approximation that reasonably reflects the key factors and which yet remains comprehensible, workable and consistent. The best approach to this is to think it through in commonsense terms, taking account of the key factors, one at a time, until we reach saturation – which doesn't take long.

Item number	A12984-1
Description	Headlamp assembly
Lead time	5 weeks
Shelf life	None
Importance	VOR item
Unit cost	£32.55
Units in pack	1
Item class	A
Sales in past 12 months	15 12 3 33 17 23 4 18 1 12 22
12-month average	15.25
6-month average	13.3
Standard deviation	9.1

Figure 4 *Data relevant to setting order point and order quantity for a headlamp assembly.*

The most obvious point at which to begin is to consider the lead time. If the lead time is 5 weeks, then the order point should be at least 5-weeks stock – based on some estimate of demand – otherwise we would be planning to run out of stock before the replenishment arrived.

But 5 weeks is a bare cover. The lead time might be extended or the demand increase above the estimate, in which case the 5 weeks would not be sufficient. For this reason, we hold some 'safety' or 'buffer stock' in addition to the 5 weeks. The formula for order point at this stage then becomes:

Order point = Expected demand during lead time + safety stock

The expected demand is ascertained either from a forecast – normally supplied by the sales or marketing departments – or from a calculated average. If this is expressed as a single monthly figure, by far the most common approach with large numbers of items, then we have:

Order point = expected monthly demand × lead time + safety stock

Now here it looks like becoming interesting. What is the next step? How do we go from here to the idea of taking into account all those other factors we mentioned earlier – item importance, unit cost, demand variability, and so on? Well, you are going to be disappointed if you expect much more because it stops here. And this is not only what I think. The foregoing formulations, and specifically the first of them, is almost universally accepted at a theoretical level – which is not to say it is always used in practice. The other factors – importance, lead time, demand variability, unit cost, and so on – do not figure explicitly in the order point formulation; rather, they tend to be incorporated through the setting of the safety stock. So how do we set safety stock?

Safety stock

The critical role of the safety stock settings is obvious. Safety stock is the main factor determining the balance between inventory investment and service to the customer. High safety stocks mean a high service but also imply a high investment in inventory, and vice versa.

Safety stock settings are also the medium through which management objectives should be reflected in the inventory control system. If the company is aggressively seeking market share and emphasizing the priority of customer service, then safety stocks might be generously set; if, conversely, the company is struggling for survival in the face of a cash flow crisis, then management in its wisdom might decide to lower the safety stock settings.

The guidelines for setting safety stocks are reasonably easy to set down, although they could well prove matters of debate. Important items, for example, might have higher safety stocks than less important items; it might be thought best to carry higher safety stocks of lower-value items to give a higher overall service from a lower total stock value. And so on. So how do we go about setting safety stocks?

The simplest approach is to express the safety stock as so many units or weeks of stock. Such calculations are easy to understand but difficult to relate to concepts like the service level we seek.

Alternatively, we could develop a mathematical model to take account of the service level we are after and the various factors we mentioned earlier, or at least some of them. The formulations that tend to be used the most rely on relating safety stock to some measure of demand variability (for example, standard deviation (SD) or mean absolute deviation (MAD)) and/or a measure of lead time variability (again SD or MAD of lead time). It is very rare to find a safety stock formulation in a real inventory situation which formally takes factors like unit cost or importance into account. The difficulty with such models lies in the general lack of understanding of them possessed by both managers and stock controllers.

A third method of setting safety stocks is to divide the inventory into classes or categories and then to set differing safety stocks for each of the categories. The ABC classification is perhaps the best known of these.

These three approaches – simple weeks of stock, using a mathematical model that relates safety stock setting to the desired service level and setting different safety stocks for various categories – are the main methods used in practice. Often, they are used in combination. In Chapter 3, we will look at the mathematics of safety stock determination in more detail and then return to the question of how a real company operating in the real world should go about setting and maintaining safety stocks. But let us now turn to the upper level of the two-level system.

Setting order quantity

The approach to setting order quantity which is to be found in the textbooks is that of the 'economic order quantity (EOQ)'. Several versions of this method have been developed to cover price breaks and the like, but the one invariably used in practice is that of Camp or Wilson, which has been around for over 50 years. This is derived in Appendix 1. Aside from this, there is no generally accepted alternative 'formula' for setting order quantity – despite the fact that the EOQ formulation itself is not widely followed.

The only other systematic approach that is used in practice is to set the order quantity in 'weeks of stock'. Typically, the idea is to order A items frequently and C items less frequently – an idea that is not at all dissimilar to the concept underlying the EOQ formulation.

These two methods are really the only formal approaches that tend to be used in the setting of order quantity (or MAX in the case of the MAX/MIN system). Clearly, any approach must take account of minimum pack sizes and the like.

Two-level systems in summary

These are by far the most widely used systems of inventory control in warehouses.

In two-level systems of inventory control, the user decides both when to order (determined by the order point) and how much (the order quantity). The order point is calculated as follows:

Order point = expected demand during lead time + safety stock

The order quantity is set either using the EOQ formulation or in terms of 'so many weeks' of stock.

The MAX/MIN system is similar to the OP/OQ approach, except that the order always tops up to the MAX. In both cases, four variables are used:

1. Forecast of demand.
2. Supply lead time.
3. Safety stock.
4. Order quantity.

The first two of these – forecast and lead time – are (or should be) best estimates of future demand and the time it takes to replenish; in this sense, they are 'guessed' rather than 'set'. But the others – safety stock and order quantity – are 'set' by the user, depending on what he is trying to achieve; they are the variables through which he fine tunes his inventory control system.

Every inventory system possesses the same four variables – forecast, lead time, safety stock and order quantity (or its equivalent). And, in every case, the safety stock and possibly the order quantity must be set by the user. The essential problem of inventory control is that people are unsure of how they should go about setting safety stocks and order quantities.

One-level systems

The concept of the ordering window

The two-level systems assume that the order on the supplier can be placed at any time – whenever the order point is reached. But there may be cases when orders can only be placed at certain points in time. These are known as the 'ordering windows'. For example, a steel mill may have a rolling cycle that closes on a certain day – so orders must be in by that day for delivery the following month. An importer of Japanese television sets may place a monthly order on his supplier either because of contractual arrangements or convenience. In such cases, the decision of 'when' to order no longer lies with the user: he now has only one decision to make when the ordering window comes around – 'how much' to order.

For a one-level system, there are two conditions that must be fulfilled:

1. Orders can be placed only at the ordering windows.
2. The item is normally ordered at every ordering window.

This second condition raises an important, although not immediately obvious, point. In the monthly order of television sets from Japan, for example, the normal practice would be to place an order for *every* model every month: there are relatively few products and each is of reasonable unit value. This is a true one-level system. But a company that imports brake linings from Japan, and which also places a monthly order on the supplier, would end up placing about 5,000 small orders each month if they used the same approach. They would therefore tend to use it only for their fast-moving 'A' lines: the others would be controlled by a two-level system – MAX/MIN or whatever – even though there would only be one time each month in which an order could be placed.

Clearly, with a one-level system with a single monthly ordering window, the average order size would be one month's stock. Also, if there was no movement during a month, then no order would normally be placed.

The top-up level

At first sight, it might seem that this one-level system is one of extreme simplicity – all we need to do every month is to order one month's stock. In fact, this is quite erroneous – a 'rule' like this is unworkable. Some type of 'balancing' departures from forecast must be included.

The simplest way to do this is to top up the stock at every ordering window to a predetermined level. For this reason, the one-level system is known as either a 'top-up' system or a 'periodic review' system.

As there is only one decision to make, only one level is needed. It is equivalent to the upper level of the two-level system; we will call it the 'order-up-to level (OUTL)'. There is no equivalent to the order point because no order trigger is needed. The operating rule for the periodic review system is:

'At every ordering window, place an order that is sufficient to top up the assets (on hand + on order − customer backorders) to the OUTL.'

But how should the OUTL be set? Consider a simple example – ordering television sets from Japan.

Suppose we order at the beginning of each month for delivery in the middle of the following month – the lead time is 6 weeks. The time between ordering windows, the 'review time', is a month, say 4 weeks. It is now the beginning of the month. If for the present we don't include any safety stock, how many weeks stock do we need on hand or on order (including this present order) – that is, what is the top-up level expressed in weeks stock? Take some time to work it out.

The answer is 10-weeks stock – sufficient to last us until the arrival of the next order after the one we are placing now. If you don't see why this is so, think about it until you do, as it is the basis of one-level systems. (But if you can't work it out for yourself, it is explained in Appendix 2.) This is the sum of the review time (4 weeks) and the lead time (6 weeks).

In practice, to allow for the fluctuations in the lead time and in the expected demand over the next 10 weeks, some safety stock is held – just as with the two-level system.

Putting these together, we have the formula for the OUTL:

OUTL = expected demand during (lead time + review time) + safety stock

If we set the safety stock at 3 weeks in our example, then the OUTL would be (6 + 4 + 3) or 13-weeks stock. If the average sales were 40 per week, then this would be 520 units.

As with the definition of order point, there is a universal agreement at the theoretical level on this formulation but a patchy use of it in practice.

One-level systems in summary

One-level or periodic review systems are used to control an item if:

- orders can only be placed at certain ordering windows and
- the item is ordered at every ordering window.

The formula for the OUTL is:

OUTL = Expected demand during (lead time + review time) + safety stock

This is similar to the upper level in the two-level system.

Materials requirements planning systems

Dependent and independent demand

In a warehouse situation, the demand for one item, in general terms, is not related to that for any other. This is not the case in a production operation. If A is made out of B and C, then the requirement for B and C relates directly to that for A. These demands are not independent of each other: the demand for B and C is dependent on the demand for A. It would therefore seem logical to work out the requirements for B and C from the upper-level requirement for A. This is what 'materials requirements planning (MRP)' attempts to do.

The logic of MRP

The logic of MRP is simple in concept but rather more involved in operation. The whole system is driven by a 'master production schedule', which is a statement of what the manufacturing operation is going to make. Depending on the nature of the operation, this schedule may be derived from customer orders, a forecast of demand or a combination of both. The schedule horizon should cover the cumulative manufacturing and purchasing lead times.

Using the master production schedule, the bills of materials and the lead times, the top-level requirements are 'exploded' to give the demands for the next level down. The process is then repeated, level by level, until the requirements at every level are ascertained. During the process, the on-hand balances, outstanding works and purchase orders are taken into account, as are parameters like safety stock and scrap allowances. As the calculations involved are very numerous, this approach to planning is normally conducted with the aid of a computer. There are numerous MRP packages available nowadays.

MRP as a solution to manufacturing

MRP packages are often offered as the computer 'solution' to manufacturing. In fact, they are the only general-purpose manufacturing packages available today.

While there are many operations that could not survive without an MRP-type package – fuel pump manufacture, for example – there are many manufacturing situations that do not need or cannot use the MRP approach to planning. Such packages are irrelevant in continuous process operations like cement or fertilizer manufacture. They are an overkill if all that is needed is a bill of materials to determine packaging requirements – as with paper tissue manufacture. In screw manufacture, with maybe 50 raw materials – mainly wire – and thousands of end products, and where the production processes are shaping one item rather than combining several, MRP planning tools are not needed. Yet it is not uncommon to find a manufacturing company that has spent a good deal of money on an MRP package that offers it little assistance in their production operations.

In conclusion, MRP packages can be of enormous benefit to many manufacturing activities but, as in any choice of systems, it is necessary to ensure that the 'solution' matches the 'need'.

Japanese approaches to control

The spirit of Just-in-Time

There is a well-developed, essentially western 'theory' of inventory control. A major part of this theory is concerned with the setting of parameters – and safety stocks in particular – to cope with uncertainty. Partly because of the emphasis on theory and partly because of the fear of running out of stock, the past endeavour in the west at the practical level has been to 'cope' with uncertainty rather than to attempt to 'control' or limit it. Most western inventory control systems hold high formal and informal buffers to cover against uncertainty. Besides the formal buffer of safety stock, we are often generous in our setting of lead times or we order before we need to. Altogether, buffer stocks account for more than 50% of the total stockholding in most warehouses – a high price to pay to cover, albeit sporadically, against high demand peaks and lead time extensions.

The essence of the approaches characterized by terms such as Just-in-Time (JIT) and Kanban has been to concentrate a great deal of attention on reducing the uncertainty surrounding both demand and supply, as well as on honing internal operations. In some cases, and specifically with high-volume assembly lines, this has led, over a period of years, to a situation where stocks previously measured in weeks or months are now found in hours or days.

It is often said that JIT has little relevance to warehouse operations with thousands of items, a multitude of suppliers spread across several countries and with lead times measured in weeks or months rather than hours. This is true if we think of JIT as things arriving just before they are required. But if we look at the spirit of JIT, which works from the basic hypothesis that any situation can be improved by the progressive identification and elimination of wasteful practices, then we have a philosophy that is as valid in a warehouse operation as it is on the Toyota assembly line.

The investment service balance

It is commonly argued that if we reduce inventory, then the service we offer to customers will also degrade. In fact, this often happens in practice; a crash program of stock reduction in response to management pressure often results in a deterioration in service. But it is wrong to conclude that this is anything like a universal truth.

If the operation is inefficient – not using its resources as well as it might – then it is possible to maintain or even increase service while at the same time reducing stock by increasing the operational efficiency. In a mine store, by simply introducing a systematic approach to inventory control where previously it had been a highly subjective exercise, a stock reduction of several hundred thousand dollars was achieved while service increased by four percentage points. A group of pirate automotive spare parts operations was able to reduce its overall inventory by 25% over a period of 18 months by simultaneously rationalizing its product range and devising a system of single-site stocking of slow-moving products throughout the group.

In both of these cases, the improved balance between investment and service came not from the reduction of uncertainty in demand or lead times, but from the smartening up of internal operations. This is a very fertile field to plough in most warehouse situations. In the first chapter, some of the inadequacies often found in current operations were pointed out: there are likely to be others.

There is considerable waste in the manner in which most of us manage and control inventory using computer-based systems. Only some of us may be able to significantly reduce the uncertainty surrounding lead time and demand, but the majority of us have a great opportunity to improve the overall inventory situation by concentrating attention on our internal systems and procedures. The difficulty is not the lack of opportunity to improve but the complacency with existing modes of operation.

In conclusion

In this chapter, we have reviewed the different types of systems that are used as a basis for the control of inventory. This has been a 'theoretical' review in that the emphasis has been on discussing the underlying logic of the systems and on deriving the relevant equations. Moreover, the systems discussed are the ones that tend to be used in most real-life inventory situations.

For most warehouse operations, the order point/order quantity (or MAX/MIN) system is appropriate. The periodic review system tends to be used primarily for those operations that can only order at certain times and, within these bounds, only for the high-value and/or fast-moving items. Materials requirements planning is normally used in the context of a multi-level manufacturing operation.

In both warehousing and manufacturing, the idea of not holding unnecessary buffers, and beyond that of working to structure the situation so that buffers are not needed because of reduced uncertainty, is a lesson that most western operations could take to heart from the Japanese.

Before we see how all this relates to the real world of stock control, let us briefly look at some of the more commonly used theoretical algorithms of control. This is the 'mathematical' aspect of stock control and, while the author feels that such methods often tend to confuse rather than assist the process of control, any manager with an interest in inventory control should be aware of what is available.

3
MATHEMATICS IN INVENTORY CONTROL SYSTEMS

Overview

Mathematics has always been associated with stock control. In this chapter, we look at the algorithms that tend to be used in practice, or at least advocated for such use. The underlying mathematics is not developed in the body of the chapter but in Appendix 1; this will enable those with a more quantitative orientation to follow the material through in a more formal manner.

It is well to be aware of the limitations of such theoretical constructs, and to distinguish between plausible claims and proven performance. As we proceed, we will attempt to highlight both the advantages and the disadvantages of each technique in the hope of leaving the reader with a balanced view of their applicability.

Systems and mathematics

The place of mathematics

After decades of theoretical development, the operations research literature is now replete with inventory control algorithms of considerable breadth of treatment and depth of mathematical reasoning. Very few of these are used to any extent in real-life operating inventory control systems. Where mathematical approaches are used, the overall tendency is to continue to use a few algorithms that have been around for a long time. In this chapter, these perennials will be reviewed.

Systems and parameters

In the last chapter, the different types of inventory control systems were detailed. In each of these, there were four parameters on which the ordering levels depended. These were:

1. Forecast.
2. Safety stock.
3. Lead time.
4. A parameter that spaced out the orders. (EOQ, MAX-MIN, time between ordering windows).

The introduction of mathematics does not mean that further systems are being considered. Rather, the mathematics is used to determine these four parameters within the context of the systems already discussed. Most of the theory is concentrated on forecasting, the setting of safety stocks and EOQ formulations. We will consider each of these in turn in the following sections.

Forecasting

Forecasting and averaging

I have met many managers who draw a distinction between 'averaging' and 'forecasting' in the sense that the first is what they do now while the second is what they would like to do in the future. The implication is that averaging is a stopgap measure, something that will keep them going until they get around to forecasting. When people think in these terms, they usually have in mind some mathematical approach to forecasting about whose details they are a little vague but which, in their perception, has the great attraction that it is able to keep up with market trends. It is a curious mixture of faith and gullibility.

In this section, the approaches that can be used to forecasting for inventory control purposes will be reviewed.

Time series analysis

Forecasting is based on the analysis of past patterns of demand, a discipline known as 'time series analysis'. The underlying assumption of any mathematical forecasting technique is that the factors at work in the past will be at work in the future. If these patterns or components, as they are called, can be identified, then they can be used to estimate or forecast the future demand.

Such approaches generally recognize four main components in a pattern of demand:

1. Trend.
2. Seasonality.
3. Cyclic.

4. Random.

The ideas of trend and seasonality are self-evident. The cycle of concern here is the business or other cycle relevant to the item under consideration, and may stretch over many years. But the random component is the most interesting. The term itself sounds quite impressive, almost scientific. But this is nothing more than a dumping ground for all those causes of variation in a demand pattern that cannot be ascribed to trend, seasonality or cycle.

The techniques of time series analysis attempt to 'decompose' the series into these four components. The textbooks tend to exemplify the approach by using cases like the retail sales in the United States over a period of several years – a long-favoured example that demonstrates significant trend, seasonality and cycles with comparatively little randomness. Unfortunately, if the major components of trend, seasonality and cycle are not the prime cause of variation from one period to the next – that is, the random component is predominant - then the mathematical techniques are not of much use in estimating the future. This is often the case with monthly sales at the item level. An example will illustrate.

Consider the sales of a toothbrush by a supermarket for a 12-month period:

23   67   34   84   25   98   24   38   71   12   87   66

These are real figures. The question of interest is how many are expected to sell next month. This is a non-seasonal item; there does not seem to be any trend and it is a fair assumption that the month-by-month variation owes little to the current business cycle.

The reason sales vary from one month to the next is not known to us – it is random market noise. In such cases, there is little advantage to be gained by using sophisticated time series analysis techniques to develop a forecast. An average would perform just as well. This is a conclusion that has wide applicability in inventory control.

Methods of forecasting for inventory control

With the present power of computers, it is technically feasible to forecast for large numbers of items using quite advanced time series analysis techniques. But because of the high random component in most patterns of demand at the item level, there is little serious suggestion that these techniques be used for forecasting for inventory control at this level. So what are the options?

The techniques that tend to be used for the most part are either the 'simple average' or the 'exponentially smoothed average'. These are by far the most common techniques in use today.

A simple average is found by adding up the past 'X' periods demand and dividing the result by 'X'.

Exponential smoothing uses the previous average and adds on each month the difference between the forecast – the old average – and the actual demand that occurs. In this way, the average is constantly 'adapting' to changes in demand. It is a simple concept and is readily seen in the following formula:

New average = old average + alpha* (latest demand − old average)

Alpha is the smoothing factor, normally set at between 0.1 and 0.2. Thus, if alpha is 0.1, we have:

New average = old average + 0.1* (latest demand − old average)

So, if the old average was 25 and the latest demand turned out to be 30, the new average would be 25.5 (25 + 0.1* [30 − 25]).

The formula for exponential smoothing is more usually expressed as:

New average = (1 − alpha)*(old average) + alpha* (latest demand)

or, if ALPHA is 0.1:

New average = 0.9* (old average) + 0.1* (latest demand)

It is also possible to express the exponentially smoothed average in terms of past demands, as is illustrated in Appendix 1. When this is done, the weights given to more recent periods are seen to be higher than the weights given to periods further back in history. With alpha set at 0.1, for example, the weights are:

0.100, 0.090, 0.081, 0.073, 0.066, 0.059, ...

with the first applying to the most recent period.

Neither the simple average nor the exponentially smoothed average accounts for trend (despite the claims frequently made for the 'adaptability' of exponential smoothing). To take trend into consideration, the appropriate extensions have to be applied: for averaging this is linear regression and for exponential smoothing it is double-exponential smoothing. But these two methods do not account for seasonality.

Strangely enough, even though it would be a simple enough task to program the algorithms for trend and seasonality into most computers, little attempt has been made to do this in most computer-based stock control systems. The simple truth is that most computer-based inventory control systems and packages:

- Use either averaging or exponential smoothing (if they make any formal attempt at forecasting at all).
- Do not account for trend.
- Have no formal method of either determining or using seasonality.

These bald facts often conflict with management's perception of its operations. It is not uncommon for people to think that they have a sophisticated forecasting system simply because they use exponential smoothing. This is a misconception.

In the following sections, we will look at some aspects of forecasting in a little more detail. Our objective is not to describe the mathematics underlying the techniques – this is left to Appendix 1 – but to give a feeling for the relevance of the techniques in the practice of stock control.

Exponential smoothing

Exponential smoothing provides an excellent example of a solution to a problem that is no better than the alternatives, but which is accepted because it sounds plausible and the manager does not really understand it. The prime requirement of a forecasting technique in the eyes of management is that it keeps up with developing trends in the marketplace. The appeal of exponential smoothing is that it is often presented as being able to do this:

'Exponential smoothing gives greater weight to more recent demands. Therefore, it adapts to changing trends in the marketplace better than a method that gives equal weight to all months considered (like an average).'

Now this sounds quite appealing, but it is nonsense. Despite the fact that the first sentence is true, the conclusion presented in the second sentence does not follow from it; it is a very good example of a plausible untruth. This is demonstrated in Figs 5 and 6, where the relative performances of averaging and exponential smoothing are compared quantitatively.

The following conclusions can be drawn about exponential smoothing for inventory control:

- The techniques of moving average and exponential smoothing perform similarly. Both can be made 'adaptable' – for demand patterns displaying trend – or 'stable' – for patterns with much variation but little real trend – simply by varying the number of months or the alpha factor. There is no superior adaptability with exponential smoothing.

- It is highly debatable whether forecasts should be as adaptable as people seem to think they want. Many managers tend to think that the marketplace is characterized by 'developing trends'. The truth of the matter is that even when such trends are evident in overall patterns of demand – at the product group or business segment level, for example – they are less in evidence and difficult to pick up at the item level. Usually, they are overwhelmed by the random component. An 'adaptable' method jumps up and down from one month to the next, not because of 'trend', but primarily in response to these random fluctuations. This is illustrated in Fig 7. It is very disruptive in inventory control to have forecasts changing so drastically in response to noise.

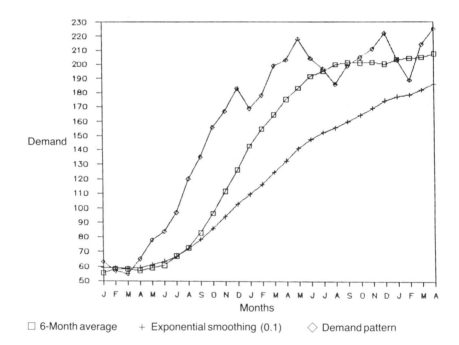

□ 6-Month average   + Exponential smoothing (0.1)   ◇ Demand pattern

Figure 5   *A 6-month average adapts to the new level of demand more quickly than exponential smoothing with a smoothing constant of 0.1. But both approaches can be made more 'adaptable' simply by increasing the alpha factor (for exponential smoothing) or reducing the number of months (for an average).*

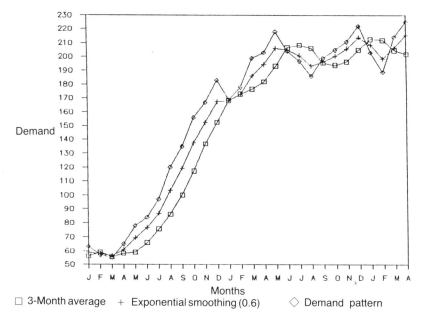

Figure 6 *Here, exponential smoothing with alpha set at 0.5 is slightly quicker in adapting than a 3-month average. A 2-month average would have proved more adaptable than either of these methods.*

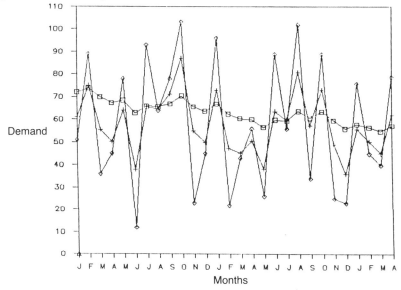

Figure 7 *With a highly variable pattern of demand, the highly adaptable method (exponential smoothing with alpha = 0.5) performs badly as it responds to random fluctuations in demand. A less adaptable method (exponential smoothing with alpha = 0.1) keeps to the centre of the demand pattern. The same conclusion would apply if simple averages were used.*

What would be very desirable is a method of forecasting that works out for itself whether the pattern is 'trending' or 'variable' and then applies a forecasting method appropriate to each case. This shouldn't be too difficult, since the past history indicates the nature of the pattern. Adaptive smoothing – a modified form of exponential smoothing in which the alpha factor depends on the extent of variation of past demand – is perhaps the best-known such technique. It sounds an extremely attractive technique, although it does not perform as well as might be anticipated in practice, and so is not widely used.

Exponential smoothing was introduced into computers as a method of forecasting, not because it was a superior method of estimating the future, but simply because it required less data to be stored at the item level. In those days, the late 1950s, data storage was expensive and data minimization was a priority in systems design. Nowadays, data storage is inexpensive yet exponential smoothing has survived in many computer packages of inventory control. Why is this?

Perhaps it is easier to sell a package using exponential smoothing. If you, as a manager, have had grievous problems with forecasting in the past and are looking for a new stock control computer package, would you be impressed with a package in which the power of the computer is used to find nothing more than an average? Or would you be more inspired by an alternative package which places considerable emphasis on the 'adaptability' claims of exponential smoothing?

Where exponential smoothing is used, the most common value suggested for the alpha setting is 0.1. It might be concluded that this suggestion is the result of a multitude of simulation exercises geared towards the determination of an 'optimal' alpha value. But this is not the case. The person responsible for introducing exponential smoothing in the first place, R. G. Brown, explains the true reason:

> 'There is a reason why 0.1 was originally chosen in exponential smoothing. On the old IBM type 602A multiplying punch, one didn't want to multiply two numbers if possible; it was too slow. But it is easy to multiply by 0.1. Move the wires over one, hold in the plus and add. That's where the magic smoothing constant of 0.1 really came from.'

Management ignorance and model plausibility are a dangerous combination.

Trend

The adaptability of both a simple average or an exponentially smoothed average is achieved by heavy weighting of recent months. In this way, the calculated average rises or falls quite quickly if there is a sustained change in demand. This adaptability is quite different from the concept of 'trend' – even though we used the term rather loosely in the last section as in talking of 'developing trends' in the marketplace.

Trend in the mathematical sense refers to the underlying rise or fall in a pattern of demand. Using either regression or double-exponential smoothing on past data, it is possible to calculate a trend line which can be projected to forecast future demand. These techniques are discussed in more detail in Appendix 1. Neither technique is much used in real inventory control systems at present, but, if desired, either could be readily programmed into most computers.

Seasonality

Many inventories contain seasonal items. Some are totally seasonal, like summer clothes, while others sell throughout the year but more in one period than another. Despite this, few practical inventory systems take formal account of seasonality; it is extremely rare to find an operating system that calculates and/or uses seasonality for inventory control purposes. The more common approach is for seasonality to be input manually and informally into the ordering considerations.

The reason for this is simple enough. While seasonality does exist in inventory items and while there are several algorithms by which seasonality could be taken into account, putting all this together in a system that takes formal and meaningful account of seasonality for large numbers of items is not at all easy. For example, while seasonality tends to be most commonly expressed in terms of months, many seasonal factors – like Easter or when the rains come or the first cold spell of winter – cannot be identified with specific months on a year-by-year basis. The random fluctuations in demand patterns coupled with factors like promotions often provide further distortions that obscure seasonal patterns.

The mechanics of accounting for seasonality are relatively simple. The essential idea is the seasonal index. Typically, a seasonal profile may be developed which consists of 12 seasonal indices which indicate the relative proportion of sales expected in each of the months. These are then combined with the average to give month-by-month forecasts of demand.

Indices may be derived from an analysis of the *sales for an item* over the past several years. Generally speaking, it is desirable to have 3-years demand history. The method of calculating the indices is quite simple and intuitive: add up the January sales in all 3 years to obtain the January figure, and so on – see Appendix 1. If the random component is greater than the seasonal component, then it may be difficult to derive reliable indices in this manner. Unfortunately, this is almost universally the case, unless the seasonality is extreme or the item is a very fast mover. Seasonal indices calculated for the sales of a high-selling motor vehicle are usually quite reliable and useful, but the indices calculated for the after-market sales of a particular body panel for that same vehicle – even knowing that panels sell more in winter – may be less so.

Alternatively, the indices may be generated by applying the historical overall group sales pattern by month to each item in the group. This approach tends to be most useful when a whole range of products is seasonal – like ice-cream.

These are the prime methods by which seasonal indices may be calculated. While the precise manner in which they are incorporated into the forecast depends on the underlying forecasting technique being used, this is intuitively clear and will not be pursued further here, although it is discussed in Appendix 1. The basic difficulty is in calculating the indices, not in using them.

Forecasting in summary

Most inventory control systems that are not provided with item forecasts by a sales or marketing department use exponential smoothing or averaging to calculate a forecast of future demand. Few of these take formal account of either trend or seasonality.

There is no real difficulty in programming the algorithms for trend or seasonality or both into a modern computer. Their general absence is a reflection of the difficulties involved in using these algorithms for practical stock control purposes. It may also be indicative of a general management ignorance of forecasting techniques coupled in some cases with an enhanced but erroneous perception of the power of their existing systems.

The final word on forecasting comes from George Plossl:

> 'Forecasting in business is like sex in society: we have to have it; we can't get along without it; everyone is doing it one way or another, but nobody is sure he is doing it the best way.'

Safety stocks
The role of safety stock

Safety stock is held to cover against an extension of lead time or the possibility that actual demand is greater than forecast. If the goods are delivered within the lead time, or if the demand is less than expected, then the safety stock is not only not needed, it is now in excess of requirements. So, in any real situation at any point in time, holding safety stock for some items ensures that we can meet the demands of the customer, while for others it means that we have more stock than we need to do this. Most traditional western manufacturing and warehousing operations carry generous safety stocks. Consider an example.

An item is ordered from overseas with a lead time of 12 weeks. Control is exercised using a MAX/MIN approach with the MIN set at 16 weeks and the MAX at 20 weeks. Do you think that these are reasonable settings for such an item?

Most of us would not think that these levels were excessively high. I recently worked in a company with a 3-month lead time for most products; they set the MIN at 22 weeks and the MAX at 26 weeks.

Here's a simple question. What proportion of the stockholding in this example is safety stock? We know the safety stock is 4-weeks stock, but what percentage is this of the expected average stockholding? Surely this is a pertinent question – managers should be interested in the proportion of their total stock that they plan to hold to cover against high demands of lead time extensions. See if you can work this out before you proceed further. My experience in seminars is that most managers cannot do this, which must say something about the average manager's ability to think through the implications of inventory management decisions in quantitative terms.

In Appendix 3, the proportion of safety stock in this situation is calculated: it is 67%. Two-thirds of the total stock is held to guard against the supplier not delivering on time or unusually high demands. This is not atypical. As we saw earlier, in addition to these formal safety stock holdings, many operations may hold significant informal buffers as well.

Given the fact that it represents such a large proportion of our total stock, it follows that the manner in which safety stock is set is deserving of consideration. In fact, it is the key parameter through which management control should be expressed. The options for setting safety stock are discussed in the following section; a more comprehensive mathematical treatment can be found in Appendix 1.

Approaches to setting safety stock

At the simplest level, safety stock may be set in units or in weeks of stock. There is a problem inherent in this approach. Consider the two patterns of demand shown in Fig 8.

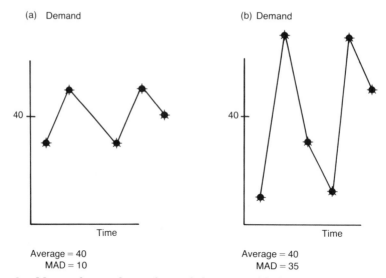

Figure 8   *More safety stock may be needed to cope with the more variable demand pattern in (b).*

Both of these patterns have an average of 40, but A has a reasonable pattern of demand while B is much more variable. Clearly, a greater safety stock would be needed in B than in A if we wished to offer the same type of service to the customer in both cases. But if safety stock is set in terms of the average, say 4-weeks stock, then this would yield the same safety stock for both cases (40 units). What is needed is a method of setting safety stocks which takes account of the variation in demand and so would give a higher safety stock in case B.

Such approaches exist and the most common are those based on the standard deviation (SD) or the mean absolute deviation (MAD). Both of these measure the extent of variation about the average. In the example, while both A and B have the same average demand (40), the variation about this average figure is reflected in the MADs - 10 for A and 35 for B. By using SD or MAD, safety stocks may be set to reflect the variability of the demand patterns.

The service level approach

This line of argument can be taken even further. Consider an item with a 4-week lead time. How much safety stock is needed? Would 3 weeks be sufficient or should it be 2 standard deviations?

Here is a fundamental problem. The prime parameter in any inventory control situation is the safety stock. It should be the parameter through which the management objectives of the company are to be reflected. Yet managers have no clear ideas about what it should be. It is not at all easy for them to relate this safety stock setting – whether it be in terms of units, averages or standard deviations – to the objectives that they are concerned with – primarily investment and service. As a consultant, I am constantly being asked by managers questions like 'How should we set safety stocks to achieve a 95% service level?'

Clearly, service level is related to safety stock: the higher the safety stock, the greater the service. Is it not possible to develop a mathematical model to relate the two? The answer is 'yes' and in fact several such models exist. Some of these models are outlined in Appendix 1. The basic idea is that the user plugs in a 'desired service level' and the computer calculates the safety stock required to give that service. This is clearly the answer to a manager's prayer.

But the question the manager should be asking is 'Should I use such a model?' The answer depends on the extent of the validity of the model; that is, how likely it is that by setting a desired service level of 95% he will achieve a 95% service in practice. He also needs to consider the cost implications of such an approach – How much stock is he likely to end up with? But, while some managers attempt to follow through on the second question using simple simulations, comparatively few even attempt to gain some feeling for the sensitivities relating to the first. Nor is it easy to do so, as the type of simulation required is rather more complex than that needed to simulate the cost implications of an inventory control policy.

There are advantages and disadvantages in using such models. The prime advantage is that they quantify a relationship that is difficult to estimate by other means – that is, the relationship between safety stock and service level. The user does not now have to think about what the safety stock settings should be: all he has to do is to set the desired service level. Moreover, as these models are all based on the SD or MAD approach to setting safety stocks, the safety stock setting reflects the variation in demand, something that an approach based on 'weeks of stock' would not do.

What are the disadvantages? The most obvious is that the actual service level may not come out to be the one set. The model takes account of only certain factors and there are others that could impact on service. Erratic lead time performance by the supplier (if this is not taken account of by the model), an ordering person not following the system suggestions (quite common with such algorithms) or unusual demand patterns, plus a multitude of other factors, might not result in a 95% service in practice despite setting it as the desired service level. In such situations, it is not possible to know what the real service level will turn out to be, which rather nullifies the point of the exercise. In one operation in which the management objective was a 92% service, the stock controller set the desired service level at 97% to make sure he delivered.

Perhaps the most significant problem with the service level approach arises from the fact that slow-moving items display a greater variance to mean ratio than faster-moving items. In simple terms, if the average is 10,000 per month, then the greatest demand in a 12-month period might be 15,000 and the least 6,000; if the average were 10 per month, then the monthly figure could well rise to 20 or drop to zero in a 12-month period. This means that the MAD or SD is relatively higher compared with the average for slow-moving items. Some authors contend that the SD rises in proportion to the square root of the average, so that if the average increases 4-fold, the SD only doubles. Such precise formulations are highly debatable but they do illustrate the point. Using the service level approach for slow-moving items means that the system will accumulate more stock relative to sales in the slow-moving areas. Given the problems most of us have with slow-moving stock, this might not be the wisest approach to follow

It should be noted that even the staunchest proponents of the service level approach do not advocate its use for items with patterns of demand such as the following:

$$0 \quad 0 \quad 0 \quad 0 \quad 1 \quad 0 \quad 0 \quad 0 \quad 0 \quad 0 \quad 1 \quad 0$$

Yet it is precisely with these types of items that most of us have the greatest difficulty in establishing an appropriate balance between investment and service. It is here that we really need some help in the setting of safety stocks, but the theory has little to offer – nor, to be fair, has any other approach. This is the really difficult area of inventory control and the best way to get control is to concentrate continuing attention on it – to decide on a simple but definite approach, implement it and refine it in the light of operating results.

Economic order quantity formulation

The nature of the problem

Of the four parameters in an OP/OQ system, lead time and forecast are 'estimated' from existing data or market and supply conditions while the others, safety stock and order quantity, must be 'set'; that is, the user must decide what they are to be in the light of what he is attempting to do in the marketplace and of his methods of replenishment.

As with safety stock, the problem the user faces is how should he go about setting order quantities. If an item sells at a rate of 120 per month and costs £5.87, and has a minimum order quantity of 1, how many should he order when he orders it? Clearly, to reorder every time he makes a sale may be too frequent, just as to place a single order for a year's supply may give him too much stock. A balance needs to be struck between the amount of work, or more precisely the cost, of placing an order and the cost of holding the stock. How is this to be done?

In practice, people often avoid the question by using simplistic approaches to the setting of order quantities. One approach to setting the parameters in a MAX/MIN system, which seems quite widespread in both manual and computer-based systems, is to set MIN as double the lead time and MAX as double the MIN. An item with a 3-week lead time could well end up with a MIN of 6 weeks and a MAX of 12 weeks. There is no logic in this approach, although it spares the user the problem of thinking it through in detail. At the very least, it is consistent.

The theoretical approach to a formal balancing of the ordering cost and the holding cost to determine the upper level in a two-level system is the economic order quantity formulation (EOQ) mentioned earlier. This is derived in Appendix 1.

Extent of usage of EOQ

The EOQ formula is not widely used (although the calculation is to be found in many systems). One reason for this might be that it aims to minimize a variable – total inventory cost – that is not readily measured, or at least is not of great interest to management. It is very difficult to convince a managing director that a particular inventory policy is minimizing 'total cost' if, for example, the stock valuation is rising.

But the prime reason why people do not use EOQ is the common reservation that people at the sharp end of control have about using something they do not understand. They therefore use it if it 'seems reasonable' but overrule it if it does not.

Lead time

Mathematics and lead time determination

Most companies have problems with lead time variability. Safety stock is carried in part, and often in large part, because of the lead time unreliability. Can mathematics assist in this matter?

Given its critical role, there is comparatively little attention paid to lead time in the literature of inventory control. This is not due to a failure to recognize its importance; rather, it indicates an admission that the problem of lead time is not 'mathematically tractable' – that is, it does not lend itself to mathematical analysis. This is true of many of the more important aspects of most situations.

The only approach that has any currency at all is that of 'lead time smoothing'. This is a method of updating lead time in the light of the latest actual lead time using exponential smoothing.(See Appendix 1 for an explanation of this technique.) One problem with this approach is that it may update the lead time too late.If a lead time goes out from 3 weeks to 10 weeks because of a strike at the supplier's premises, then the update would only occur following the receipt of the goods after the first 10-week lead time is experienced – by which time the strike has ended and the lead time has returned to normal. Lead times have a habit of experiencing sudden changes of often short duration – and it is difficult to take this into account by mechanical means. If my supplier rings me to tell me that his lead time has gone from 4 to 12 weeks, then the only way my system will be able to take this into account is if I change the lead time setting on the computer. I also need to remember to change it back when the crisis is over.

Sometimes the lead time extension feeds on itself. When my supplier told me his lead time had gone from 4 to 12 weeks, then I would have placed an order immediately – and so would everyone else he told. So he experiences a sudden surge in orders which means that his lead time goes out further. It can develop into a cycle but eventually it collapses. At this time, it is imperative that I set my lead time back to a realistic low value – as well as reviewing outstanding orders if this is possible – or I will be overstocked very soon.

Aside from this smoothing technique, there is no other way of using mathematical sophistication as far as lead time is concerned.

Sophistication in summary

The place of mathematical models in inventory control

There are a multitude of mathematical approaches to the calculation of the parameters of inventory control but few of them are ever used in practice. Those that are most used are:

- Exponential smoothing for forecasting.
- Desired service level approach to setting safety stocks.
- Economic order quantity formulations.

In addition, lead time smoothing is sometimes to be found.

When people talk of mathematical approaches to inventory control, they are for the most part thinking of these. There is very little else of the theory that is at all widely used in practical inventory control situations.

The most common way in which such algorithms become embodied in inventory control systems or packages is for them to be 'designed in' by either accountants, data processing people, or operations research personnel. Whenever a company spends time in designing an inventory control system specifically tailored for its own situation, then the exponential smoothing/desired service level/EOQ triad seem to end up being incorporated. This is true whether it is a support store for a cement factory, a steel stockist, a health food distributor or a spare parts wholesaler – which says a great deal about the 'special requirements' of different types of inventory situations.

There is a real problem, in my view, in that these three algorithms can be presented in such an appealing manner that they are hard to resist. Every manager wants an 'adaptable' approach to forecasting, to be able to set a desired service level, rather than messing around with safety stock settings, and to order the 'economic' order quantity. The result may be that these algorithms, which will work well only if they are applied with skill and understanding, tend to appeal primarily to managers with minimal knowledge of the dynamics and realities of stock control; in other words, they are used as an excuse for not thinking things through for themselves.

The rationale for using sophistication

Why use sophisticated approaches to the determination of parameters when there are simple ones available? Presumably because they take account of more factors, take greater account of existing factors or because they work better. In fact, this last statement is something of a touchstone for choosing a system: the best one is the one that works best. But there is very little quantitative evidence that sophisticated approaches work better, or are even capable of working better, than appropriate simple approaches in inventory control at the item level.

Over the years, I have mounted numerous computer simulations to compare alternative strategies of forecasting and/or inventory control. As a mathematician, I wished to assist the company employing me at the time in choosing the best technique for their particular circumstances. My *a priori* expectation was, at least initially, that the more sophisticated the technique, the better would be the result; but the outcomes of my studies did not support this thesis. Using actual item data, the finding in almost every case was that, provided the approach was sensible, there was little significant difference between a very sophisticated technique and a simple but appropriate one. A 2-month average, for example, was easily shown to be an inappropriate forecasting technique for a health food wholesaler, but it was difficult to distinguish between a 6-month average and exponential smoothing with an alpha factor of 0.1.

While simulations did not indicate any superiority of sophisticated mathematical techniques of forecasting or inventory control, neither have I found any evidence of improved performance due to the use of these techniques in practice. Indeed, the tendency where such techniques are in place is to work around them, or even to make program modifications to simplify them. There is no practical evidence that any operational superiority comes from the practical use of such sophistication.

Moreover, it is not uncommon for a sophisticated technique to have a negative effect. The method of design is often at fault – there may be little discussion with the inventory control people or the designer's appreciation of the mathematics may be imperfect. Also, implementation parameters need to be set and insufficient thought may be given to these – if the company is aiming at a 95% service, then a 95% desired service level may be input without further thought. Once the system becomes operational, the inventory control people may find the system gives suggestions that do not accord with their intuitive reasoning – often because the parameter settings are inappropriate – and they may gradually come to mistrust the system and overrule it constantly for this reason.

It is worth noting in conclusion that Japanese approaches to inventory minimization have not attempted to achieve their goals through the use of sophisticated mathematical inventory control techniques. Which reminds us that it is time to get down to reality.

In the next chapter, we will look at the problems associated with the design, implementation and operation of a stock control system.

In conclusion

In this chapter, we have spent some time reviewing the more commonly used algorithms of inventory control. The problem with using them can be simply put: while they perform neither better nor worse than appropriate simple approaches like averages and weeks of stock, they tend to mislead managers and confuse stock controllers.

4

DESIGNING AN INVENTORY CONTROL SYSTEM

Overview

The design of a stock control process seems simple enough. The first step is to divide the stock into categories or homogeneous groupings so that control is able to be exercised in a manner that targets on the specifics of each of the categories. Then control rules need to be formulated for each of the categories. It is this process that we will address in this chapter.

The problem people have in thinking about inventory control systems

The practical reality

There are a great many inventory control systems in operation. In all of them, decisions must be made to place orders on the supplier, and the systems used to determine these tend to fall into one of the following categories:

1. Formal rules are non-existent or poorly defined. It is extremely common to find computer-based stock control systems in which the ordering parameters – order points, and so on – have to be input manually on an item-by-item basis. In most such cases, the rules by which these parameters are to be set are at best vaguely defined.

It is very easy to establish if rules are in place or not. Ask the person in control to write down the rules that govern his ordering. Then give this written set of rules, plus history and balance data for a few items, to two or three people not involved in inventory control and ask them to calculate orders based on these rules. If the orders they work out are identical, or nearly so, then you have a system; if not, you don't. It is a safe bet that if such a written set of rules does not presently exist, then you do not have a systematic approach to the inventory control function.

2. The system incorporates a formal set of rules that are based on some or all of the following algorithms:
- Exponential smoothing.
- Desired level approach to setting safety stocks.
- Economic order quantity (EOQ) calculation.

There may be some variation here but the general tendency is to use these three algorithms together.

In this case, the key question is to see how closely they are followed. Again, this is easy to test. Get a suggested order report after the ordering person has been through it and see how many of the suggestions he changes. If this is more than 10%, or at the outside 15%, then there is, in most cases, something wrong with either the rules or the person doing the ordering.

3. The system has a set of more or less clearly defined rules that are simple rather than complex and typified by:
- Simple moving averages.
- Safety stock in 'weeks' or units.
- Order quantity in 'weeks' or units.

Here, the tendency is to arrive at oversimplistic approaches. A spare parts operation with a single supplier with a 3-week lead time might use a MAX/MIN system with MIN = 6 weeks and MAX = 10 weeks for all items. A mine stores operation might use an order point of twice the supply lead time and an order quantity of 3-months stock for everything. In each case, the conversion from 'weeks' to units is accomplished through the use of the average.

Most inventory operations can be classified into one or other of these categories. None of these, in the view of the author, represents a suitable approach to this most difficult of tasks. All three display the same deficiency: an inability to think through the control of inventory in a methodical and detailed manner.

Some companies have put a great deal of work into the design of their stock control systems and rules. This may occur as part of an overall system replacement exercise, or it may be a specific response to recurring stock control problems, or it may be because a manager – often a financial manager – becomes personally involved in designing a stock control system for the stock control people. In such cases, despite the time and effort involved, the result is highly predictable – they end up with a system based on the three algorithms.

One reason why such algorithms are chosen is quite simply that the approach sounds so logical and plausible. Exponential smoothing is selected on the assumption (albeit erroneous) that it 'keeps up with demand'. The idea of relating safety stocks directly to service is enormously attractive. The EOQ formulation is so widely distributed throughout the literature that it must be the logical approach to order quantity determination. So these algorithms are chosen on the grounds of *a priori* logical appeal rather than on proven practical performance. Like many theoretically 'best' solutions, they may not prove optimal in practice.

Another reason for opting for sophistication is that the design exercise is often conducted by people who have little direct knowledge of the replenishment process – they have never had to do it themselves. These are the accountants, the data processing people and the operations research workers of this world, or even outside specialist consultants like myself. These people read the textbooks and the system design exercise reflects this.

But a further reason for going along this path, one that possibly outweighs both of the other reasons, is something else. Nowadays, especially with the Japanese influence, there is, in many quarters, a preference for system simplicity. Simple rules, understood by those who have to use them, are the best rules; Kanban values are not based on standard deviations. But when we try to apply this philosophy to the determination of a set of inventory control rules – as we will see in this chapter – it is a rather vague, and unrewarding and unconvincing exercise.

People do not design simple but comprehensive systems of inventory control because it is neither easy nor satisfying to do so. It is very much like trying to make your way through a fog where, with every step, you are not even sure where you should put your foot much less if you are still heading in the right direction; such an uncertain path is far less enticing than the clearly illuminated road of the mathematical algorithms.

In this chapter, we will attempt to define the steps involved in developing an inventory control system for an operation. This is not meant to be a 'cook book' for inventory control system design; rather, it is intended to indicate lines of approach and to assist in highlighting the issues that need to be thought through. It is intended as a guide for thought rather than as a replacement for it.

As a first step, let us lay down a rather obvious principle for computer-based inventory control systems.

The computer should calculate the ordering levels

In an order point/order quantity (OP/OQ) system, there are two quite separate functions of inventory control:

1. Calculation of the ordering levels (OP/OQ, and so on).
2. Calculation of a suggested order based on these levels.

Most computer-based systems do the second calculation; rather fewer do the first.

A very common approach is for the user to (presumably) calculate the OP/OQ levels for each item and then to input these to the computer. This method is favoured because it is said that these levels must take account of the peculiar circumstances that surround each item: market knowledge is important and must be incorporated in the ordering parameters.

It is very easy to test this hypothesis. Take a few ordering levels that have been set on this basis and ask the person who does the setting to explain how he arrived at them. I have done this on many occasions and my general finding is that the levels set in this manner could be better described as inconsistent and subjective, rather than as based on subtle interpretation of market forces. Besides, as we saw earlier, it is possible to incorporate market knowledge by overruling system suggestions based on systematic OP/OQ calculations when they are seen to be inappropriate.

It is necessary to allow for the ordering levels to be based on market knowledge or estimation for some types of items – new items,for example. This provision normally impacts on only a small proportion of the inventory at any one time and can be allowed for in the context of a set of rules by which ordering levels should be set.

If we wish to control inventory using the computer, then we should begin by getting the computer to calculate the order points and similar control parameters for us. This means that we must clearly specify the rules by which this is to be done.

The steps in the process of specifying the rules of stock control

In devising a system of inventory control, there are two steps:

1. Categorize the stock into inventory control groups and subgroups.
2. For each of these groups, decide on a set of rules and parameters for control – always ensuring that allowance is made for special cases and the input of valid market knowledge.

These steps – first defining the categories and then setting the rules and parameters for each category – set up the system. But the initial settings of the parameters may not be appropriate. To monitor and update them, it is necessary to have feedback on the operation of the system. Then by a process of trial and error, the parameters – and posssibly the rules also –

are refined to yield a balance between investment and service that is in accord with the company objectives.

Categorizing the inventory

Categorizing the stock: stock and non-stock items

How should the inventory be categorized for purposes of control? The first distinction must be between 'stock' and 'non-stock' items. Stock items are carried with the intention of satisfying customer demands as they arise. Non-stock items are not normally carried but are ordered in only in response to a firm customer order. If we are to use this idea, then every item should be clearly defined as either 'stock' or 'non-stock'.

At first sight, this stock/non-stock distinction seems quite simple and straightforward. But there are a good number of factors that should, it seems, be taken into account in devising these categories. At the same time, it all needs to be kept simple and comprehensible. This conflict between complexity and comprehensibility is a fundamental one in stock system design.

We will use this stock/non-stock distinction to illustrate the difficulty. Even though we have scarcely begun to move along this path, we will soon appreciate how hard it is to formulate simple but reasonably comprehensive rules by considering the issues surrounding the stock/non-stock issue. Some of these considerations are as follows:

1. How are we to define the stock and non-stock items? Aside from discontinued and new lines, the usual approach is to base the distinction on movement. For example, we might decide to define a stock item as one that has moved in 3 or more months of the past year. Or we might base it on average movement or number of picks. Or we might decide to vary the non-stock/stock division depending on unit cost as well as movement. But we must make some decision.
2. The stock/non-stock distinction implies that the service we are aiming for with stock items is high while the service we are going for with non-stock items is 0%. If a customer orders a stock item, we hope to supply it ex stock, but if he orders a non-stock item we are deliberately opting not to have it in stock. Does the customer know this? Should he know? Do our sales or service people need to know which are stock and which are non-stock items? Should the distinction be reflected in our catalogue or price list? Given the stark nature of the stock/non-stock distinction, do we maybe need some type of in-between category? These are pertinent questions.
3. Guidelines may be needed in other areas. If, for example, a non-stock item comes in packs of 20 and a customer orders 4, is the company to

buy in 20 and be left with 16, or is the customer to be offered a polite refusal? At other times, the customer may ask for 5 and the purchasing person will order 10 to cover the next requirement or because the stock/non-stock distinction is not completely clear to him or because he thinks that this item is beginning to take off. It is very easy for inventory to build up in the 'non-stock' area, even though in theory there should be none.

4. Are the items to be automatically reclassified by computer at the end of each month depending on movement, cost, and so on, as stock or non-stock, or is the classification to be done by means of a periodic review by the sales people? If the recalculation is automatic, this will mean that, at the end of every month, some previously 'stock' items will be reclassified as 'non-stock', and vice versa. At the end of the next month, of course, some of these will change back. This may be confusing. But, if the sales people do the review, how do we know whether it reflects real market conditions or the quirks of the individual?

5. How are new items to be handled?

These types of questions are pertinent whatever system we use. So, how are we to derive our stock/non-stock categorization rules from them?

There is no simple answer to this question and the other system decisions that will be required as we proceed further. No one is going to tell us the answer, nor should we unduly rely on mathematics to work it out. We need to think about it carefully and come up with some rules that we realize will be approximations at best and which will never be able to adequately reflect all the pertinent considerations like those just listed. Moreover, there is a good deal of uncertainty involved - we won't know until we begin working with them whether we have got it right or not.

What do the stock/non-stock division rules look like in practice? Set out below is an example of such a set of rules for an automotive parts operation:

1. Stock items are defined as follows:
   A. Unit cost <£8.00: 2 or more months with movement in past 12 months.
   B. Unit cost ≥£8.00 and <£40.00: 3 or more months with movement in past 12 months.
   C. Unit cost ≥£40.00: 4 or more months with movement in past 12 months.

2. The stock/non-stock status will be recalculated every 3 months by computer using the above rules. A printout listing status of all items

as well as those that have changed status on the most recent recalculation will be supplied to sales, service and parts control.

3. New items – those with date-to-stock within the past year – will automatically be classified as stock items. (This was to allow the OP/OQ values to be set and maintained manually for such items.)

4. As far as possible, customer orders for non-stock items should be in full pack quantities.

A satellite paper merchant who received most of his supplies from the regional warehouse on a twice weekly basis used a simpler basis to define the items he carried. For him, a stock item was one that had moved in 3 or more of the past 6 months.

If you think rules like those just given are inadequate, try to come up with some more comprehensive alternatives yourself. It is well worth doing this, if only to gain a deeper appreciation of the difficulties involved.

Finally, it should be noted that these rules, and the ones that follow, must be based on an analysis of the inventory. In any attempt to divide the stock into categories – non-stock, A, B, C, and so on – it is necessary to know something about the number of items, the value of stock and the value of demand that falls into each of the categories we are considering. Yet few computer systems provide adequate tools for such analyses.

Now let us move on to further classifications of our inventory.

Further inventory classifications

The idea of classifying inventory into groups for ordering purposes is to treat each group in a different manner: there is no other reason for doing this. The most obvious bases for classification are one or more of the following:

- *Supplier*: Some suppliers may be more reliable than others and less safety stock may be needed for them: this would necessitate categorization by supplier. (The differences in lead time between different suppliers can be taken into account without having separate categories for different suppliers.)
- *Importance*: Some items may be more important than others. In the spares operation, vehicle-off-road (VOR) items may need to be carried in greater depth than other items.
- *Pareto or ABC class*: With large numbers of items, it is wise to order the A items frequently and the C items less frequently, to ensure that ordering effort is not bogged down by spending large amounts of time on a small proportion of inventory or sales value. Different levels of

safety stock may also be held for different categories. Sometimes, instead of a combined cost/movement categorization like the ABC classification, separate cost and movement categories are used.

- *Marketing or product groups*: The company may have different objectives relating to different segments of the business. In particular, the service objectives may differ from one group to another, depending on factors like competitor action, perception of the strategic importance of the group to the business as a whole, and other factors.

These are not the only possible bases for classification but they are the more common ones.

In many companies, existing categories designed for other reasons may not be adequate for inventory control purposes. Marketing groups in an automotive spares operation (exhaust, transmission, cooling, and so on), for example, are often not the best categories to use for inventory control. What then are the categories for inventory control to be?

ABC/FMS categorization

The most common divisions used for inventory control are based on factors like ABC category, movement, unit cost or some combination of these. To illustrate this type of approach, we will outline the ABC/FMS approach. This is only one of several similar approaches but it gives a flavour of what is involved. This gives at least 9 potential stock categories: in many real-life operations the number needed may be more or less than this.

The idea of the ABC classification is well known. The A category contains the small number of items that account for a high proportion of turnover (or, alternatively, of margin). Often it is said that the top 20% of items – the A items – account for 80% of sales value, but in practice there is considerable variation about these figures. In the C category, there is a great number of items but they contribute little to total sales value. The basic idea is that these categories deserve to be treated differently for inventory control purposes.

There tends to be general agreement on the idea of an ABC classification. Moreover, it is not difficult to achieve a consensus on the broad outlines of the approach that is to be used: the objective should be to aim for a higher service level for A items (because of their importance) and to order them more frequently (because of their mostly high movement).

So we could use the ABC classification as a basis for our inventory control rules. But consider the following two A items in an automotive spares operation:

R78123   Engine block   Unit cost: £681.54
12-Months demand: 0 1   0   0   0   1   0   0   0   0   1   0

Y66123   Spark plug   Unit cost: £1.65
12-Months demand:
243   301   278   365   222   390   277   311   264   299   342   274

The second of these items is easy to control, while the first is not. A monthly average or forecast, for example, is meaningful and useful for the spark plug but irrelevant for the engine block. Moreover, there is a high risk of ending up with a considerable value of obsolescent stock with the engine block while such a threat is not present with the spark plug. Maybe they deserve different treatment – even though they are both in the A category.

For this reason, it may be useful to distinguish further within the ABC categories on the basis of movement. Each of the ABC categories, for example, might be further subdivided on the basis of movement into fast (F), medium (M) and slow (S) subcategories. The FMS classification could be on the basis of average movement or alternatively on the number of months with movement or the number of picks.

The inventory control categories vary considerably from one operation to another. But the basic idea of splitting off both the high-selling and low-selling items for special treatment, and, within the categories so formed, refining the process further by taking account of movement, is of wide applicability. It is not the only possible approach, nor is it either clever or precise. But it is much better than not using any formal classifications at all.

An example

The analysis presented in this section is of the support store for a fertilizer factory. Insurance items accounting for about one-third of total inventory are excluded.

First, let us look at the stock/non-stock division based on the criteria of 2 or more movements in the past 12 months. We can see from Table 1 that with this stock/non-stock division about 12% of usage value (that is yearly usage multiplied by unit cost) would come from the non-stock items – but they account for 34% of the total items. Non-stock items, it must be remembered, are those that have moved once or not at all in the past year.

Table 1 *Stock/non-stock division*

Item	% Usage value	% of items	% of stock value
Non-stock	12	34	17
Stock	88	66	83

After trying out a number of different stock/non-stock divisions and looking at tables for each, the managers of the company finally settled on the ones shown in Table 1 as being most in line with what they were trying to achieve in the support operation. It is perhaps relevant to note that at the time there was a considerable problem with obsolescent stock.

They then did the same type of analysis for the ABC classification, varying the dividing lines between these categories until they came up with divisions that looked meaningful. These categories apply to stock items only and so represent 88% of total usage value, 66% of total lines and 83% of total stock value. The final result is shown in Table 2.

Table 2 *ABC divisions.*

Item	% Usage value	% of items	% of stock value
A	54	11	31
B	26	23	22
C	8	32	30
Non-stock	12	34	17

They then went on to consider the FMS divisions for each of the A, B and C categories. We will not pursue this here, but the basic logic is the same as was used in settling on the stock/non-stock and ABC categories.

There is no standard method of deciding where the divisions for stock/non-stock ABC and FMS categories should lie. Ultimately, these decisions must be made by the company – preferably using analyses like those given here. Sometimes figures are quoted – A is the top 80% of sales, B the next 10% and C the last 10% – but these cannot be expected to apply equally to all types of operations and at best provide a starting point for the analysis.

Finally, while we have talked of ABC/FMS – 9 categories in all – quite clearly the number of divisions along each of these dimensions could be less or more than three.

Determining rules and parameters for the ordering categories

Ordering levels

Having decided on the categories by which stock is to be controlled, the next question is to decide on what rules and parameters to use for each of the categories so formed.

In most cases, the basic system required will be a two-level system. Suppose we use an OP/OQ approach. There are some groups of items that deserve special consideration.

In many stores with large numbers of items, the predominant problem lies in the huge proportion of slow-moving parts. These may be variously defined but it is not unusual to find that well over 50% of all items have 4 or fewer months with movements in the past year. In one heavy equipment spares operation, 80% of items with a positive stock balance had one or zero movements in the past year, but this is exceptional. For such items an average is rather a meaningless concept.

Consider the following pattern of demand:

0 1 0 0 0 0 1 0 0 1 0 0

How should the OP/OQ be set for such an item?

One approach, if the lead time is a few months or less, would be to hold one in stock and to order another when this is sold: the chances are that the replenishment would arrive before the next demand occurs. To achieve this, the OP/OQ settings would be:

OP = 1
OQ = 1

These settings assume that the reordering rule is to suggest an order when the assets (on hand + on order − customer backorders) fall below order point, as opposed to reaching order point. For slow-moving items, the distinction between 'less than' and 'less than or equal to' is a very important one.

Another special type of item to be found in most inventories is new items. The difficulty with new items lies in estimating future demand in the absence of any established historical pattern. It is further complicated by other factors: a health food wholesaler sells 1,500 packs of a new vitamin supplement in the first month – but this is just to stock up the shelves of his customers. Once the pattern is established, his sales may average out at 350 per month.

There is a very considerable tendency to overstock on new items. Initial provisioning for new airplane engines by major airlines during the late 1970s and early 1980s is a case in point: generous monetary allocations coupled with enhanced operating characteristics of the engine components meant that there was a glut of engine parts throughout the industry, amounting to tens of millions of pounds.

There are several initial provisioning and new part algorithms developed primarily to provide high-level support to aircraft and defence equipment in the field – I developed such a model myself for an aerospace company recently. The mathematics are fine, but there is no clear way of establishing factors like MTBF – 'mean time between failures' – and similar parameters on which such models depend. It's like having a new car with no petrol – which is not all that better than walking.

But there is little use of such models in most commercial stores. For most of us, the determination of the OP/OQ values for new products involves a good deal of guesswork and little by way of mathematical algorithms. Given the tendency to overstock on new items, it is well to exert budgetary control on the process by specifying how much can be spent on this activity in a period or on a specific initial provisioning exercise for a new model or piece of equipment. The OP/OQ values would then be set and maintained manually until a pattern of demand was established – perhaps after several months.

In the discussions thus far in this chapter, we have looked at the stock/non-stock distinction, the ABC/FMS basis of categorization, and the rules of control for slow-moving and new items. In many operations, the real control difficulties centre on these questions – not on whether we use exponential smoothing or averaging to forecast future demand. None of these issues required us to even consider the mathematical algorithms normally seen to be at the heart of the stock control process. But now we come to the 'normal' items in the product range, which encompasses everything except slow-moving and new items. These are not insignificant – they normally account for 80% or more of the sales in most operations. It is just that they are much easier to control.

For these, the standard order point equation is applicable:

Order point = forecast demand during lead time + safety stock

The system questions that need to be answered here relate to the methods of determining the following variables:

- *Forecasting*: The common choices are exponential smoothing or simple averaging. If desired, these may be readily extended to allow for trend and/or seasonality.

- *Safety stock setting*: The choices are 'weeks of stock', using mean absolute deviation or standard deviation, or employing the 'desired service level' approach. The system must also allow for higher safety stocks for longer lead times.
- *Order quantity calculation*: The choices are 'weeks of stock' or the EOQ formulation.

These alternatives represent the vast majority of approaches that are to be found in use today. The choice of method may well vary depending on the nature of the stock operation, and it may also change from one category to another. But the important thing is to make a choice, rather than to assume that the more mathematical approach is the most appropriate. The best way to do this is to mount some simple simulations to compare the relative benefits of alternative approaches and different parameter settings to these three questions. This is not difficult using PC spreadsheets.

An example

The various options as far as the rules are concerned may be illustrated with a single source spare parts operation which used a simplified ABC/FMS categorization. In this case, there were only two categories along each dimension. The inventory was divided into A (80% of stock sales) and B (20% of stock sales), and each of these was further divided into fast (more than 3 months of movement in the past 6 months) and slow (2 or 3 months of movement in the past 6 months). Items with 0 or 1 months of movement in the past 6 months were automatically classified as non-stock.

AF – Fast-moving A items
AS – Slow-moving A items
BF – Fast-moving B items
BS – Slow-moving B items

When the operation was analysed and the categories decided upon, the results shown in Table 3 were found.

Table 3 *ABC/FMS divisions.*

Item	% of items	% of stock value	% of sales value
AF	14	38	57
AS	22	35	20
BF	28	12	15
BS	36	15	8

For the AF and BF categories, traditional order point type equations were used with the safety stock and order size determining parameters being set separately for each category. A 6-month average was used for forecasting, and both safety stock and order quantity were set in terms of weeks of stock. Despite the fact that these two categories together account for just under three-quarters of the total sales, they are not normally too difficult to control; an understock or overstock position becomes quickly evident and can be readily remedied by changing the pattern of ordering slightly. When the water is flowing, it is easy to control the volume.

The AS category is a potential problem area – it is very easy to be left with slow-moving, high-value items. Consequently, tight control is needed. But there is another factor – demands, when they occur, are likely to be for one item at a time. In this case, where typical lead times were under two months, the following OP/OQ values were used:

OP = 1
OQ = 1

The fourth category we are considering, BS, holds a large number of items that are characterized by both low value and slow movement. They do not constitute a high proportion of the total stock value. So the company decided to attempt to enhance the overall service at low cost by stocking this area more deeply than the AS category. So they settled on the following OP/OQ settings.

OP = Average demand in months with demand
OQ = Average demand in months with demand

In this operation, the OP/OQ values for new items were set and maintained manually for the first 6 months, at which point the automatic calculation system took over. Non-stock items had OP and OQ values set to zero.

This example gives a feeling for the manner in which different categories of stock might be handled by the use of different sets of categories. The basis of the categorization and the rules and numbers used were worked out initially using some simple simulations, and thereafter refined by close attention to how they performed in practice. At the end of each month, the stock valuations and service achieved were presented by the foregoing categories, which made the fine-tuning process possible.

This example also illustrates the fact that the end product of a thorough approach to the stock control system determination issue may well look quite trivial and oversimplistic. There are two reasons for this:

1. It is extremely difficult to work with more than a few categories – a maximum of around 10 – simply because of the confusion caused in the minds of managers and users by such a proliferation. The computer of course could work with any number; the limit is imposed by comprehension, not by a lack of computer power.
2. If we work from the baseline of using simple averages and weeks of stock in our systems, unless we can demonstrate some superiority of an alternative, more complex approach through the use of simple simulation, then we will usually end up with simple systems.

I quite vividly remember one of my early simulation projects to assist a textile company in forecasting product demand. This took months, partly because I was finding my way and partly because some quite sophisticated algorithms were modelled. At the end of the project, I presented a report to the assembled executive management group. In effect, it said that the best method of forecasting, based on months of simulation, was to use a 6-month average. It was a finding that was not well received, but it was nonetheless a valid conclusion from the data.

Demand fluctuations

In using a simple moving average or an exponentially smoothed average to estimate future demand for other than slow-moving items, a company may choose, by default, to ignore trend and seasonality in its formal systems. This tends to be done, not so much because of the technical problems associated with defining and programming a forecasting technique to take account of trend and seasonality, but because of the difficulty of reliably picking up these components at the item level. For most items, most of the time, the demand fluctuations from one month to the next seem largely random: if there is either trend or seasonality present, it tends to be obscured by the random component of the demand pattern. Like most patterns, they are easy to pick up after they have occurred, but extremely difficult to predict in advance.

There is also a further point here for those items that do in fact have an underlying trend. With lead times of up to 2 to 3 months coupled with reasonably frequent opportunities to reorder, an inventory control system using a 6-month average to estimate future demand is able to accommodate quite well to a rising or falling trend that is not too severe. Even an average increase of around 8% per month, which is equivalent to a doubling of demand every year, does not cause as many problems as one might think – the random fluctuations from one month to the next are likely to be much greater than this.

If desired, it is possible to generate messages for display on the suggested reorder report based on demand fluctuations – 'high demand', 'low demand', 'possible trend', 'possible seasonality', and the like. This seems an attractive option at first sight – pointing out to the ordering person items that may need special consideration – but it tends to perform in a rather mediocre fashion in practice. It is not easy to set the parameters that cause these messages to be produced in a consistent and meaningful manner – even defining high demand in a reliable manner is problematical – nor, partly because of the unreliability of the messages, is it easy to get people to react to them sensibly. But it is well worth a try.

In conclusion

While the steps involved in devising a stock control system are simple enough – categorize the stock and develop suitable rules for each category – it is not at all easy to do this in practice. This is more the case if we decide not to use *a priori* approaches like the desired service level algorithms.

The prime problem is a lack of clear guidelines or a framework within which to devise these rules. But, as we shall see in the next chapter, the only way to do this properly is to put in place our best guess at a set of rules and then to refine them by means of feedback from the system operation.

5

MANAGING THE STOCK CONTROL SYSTEM IN OPERATION

Overview

It is easier to design a stock system than to effectively manage its operation. Because of the impossibility of knowing precisely how a system will perform in practice, we need feedback on the operation so that we can fine tune the parameters of control. So the first requirement for managing an operating stock system is to have adequate feedback, a concept we explore in the first section of this chapter.

The actual decisions that control stock are made by the stock controllers. They know much more about the detail of the stock control process than any manager does. Effective stock control involves input from both managers and controllers, and the nature of the interaction between them is a key determinant of success. In the second section of the chapter, we look at the manager/stock controller relationship.

Finally, in this chapter, we look at the managing of the supplier relationship. In this, we can learn from the experience of JIT which has done so much to harmonize and integrate the supply chain.

These are three areas that a manager needs to concentrate on in a continuous fashion if he wishes to ensure optimal ongoing operation of his stock system.

Fine tuning the system by the use of feedback

The importance of feedback

The design of a stock control system involves two steps:

1. Categorizing the total stock into groups for purposes of control (non-Stock, A, B, C, and so on).

2. Devising rules and numbers for order parameter calculation (setting order points, and so on).

In the previous chapter, we saw how to go about settling on these categories and rules. But this initial design exercise is the easy part – the difficult thing is to make it work in such a way that the best possible results are achieved. And to do this, we need some feedback, not only on operating results, but also on the manner in which the system is being used. There are two reasons for this:

1. We may not – indeed, we are extremely unlikely to – get the rules and numbers set appropriately at the design stage. This is true if we use simulations because these are imperfect tools. But it is much more likely to be the case if we have not attempted to simulate or if we have relied on algorithm logic – for example, setting a 95% desired service level because this happens to be the company targetted service. So we need feedback to refine our approach.
2. In practice, we must expect some ordering suggestions based on our rules to be overridden by the stock controller in the light of his market knowledge; otherwise, there is no point in having him even look at the suggestions. But how many or what proportion does he overrule? Are all or at least most of the overrulings based realistically on knowledge of what is happening in the market, or is there a high subjective element here, with the stock controller overriding the suggestions because he doesn't agree with them or because he doesn't like the system? To explore such questions we need feedback.

What we are after is something really simple – a set of rules of ordering that sensibly reflect our objectives and which are followed by the stock controller, except when he knows that the expected future demand will differ from that experienced in the past. But this is not easy to achieve.

The absolute necessity of feedback by stock control category

The first and most obvious type of feedback is that relating to the objectives of the stock control process – that is, investment in stock and the service afforded to customers. There are very few companies that do not capture investment on a monthly basis and indeed this is invariably a figure of high visibility – witness the reactions when it goes over budget. Those companies that measure service also keep the performance of this dimension constantly before their eyes.

So it would seem that companies that measure stock valuation and service achieved, and react to negative variances, are well served in terms of feedback. Unfortunately, this is not the case. The ultimate test of any efficient feedback system is the ability to profit from mistakes; according to this criterion, most inventory feedback systems in use today would fail miserably. The prime use made of the limited inventory feedback systems in use in many companies is to spur a precipitous and ill-conceived management reaction to a developed crisis, rather than as a learning tool to hone and refine a system to prevent or at least minimize the onset of such events.

The idea of feedback is fundamental to control theory and is well known in the form of the 'Deming Cycle' associated with quality improvement. But there is a fundamental truth that needs to be appreciated – the importance of the feedback being in terms of the categories of control. This is readily appreciated using an example.

Suppose there is a sales manager with 17 salesmen in a region. Further suppose that all the feedback he receives from his computer is the total sales for the region – he receives no figures at all on the performance of individual salesmen. Clearly, this is not very satisfactory because he is missing precisely the information he needs to improve his group's performance and so to achieve his sales goals. In reality, the importance of ensuring feedback on the performances of individual salesmen is so obvious that it is incorporated in most sales analysis systems.

But such finesse is not evident in most stock control feedback systems. A company uses four stock control categories – non-stock, A, B and C. It has a management objective of achieving a stockturn of 4 and a 95% service to customers. Every month, the company measures both stockturn and service for the warehouse. But this approach to feedback, which is extremely common, is rather like the company reporting regional sales without individual performances – it is lacking in the precision needed for control. There are all sorts of questions that come out of this that are deserving of consideration, if not of answers. Here are some of them:

- What are the stockturn and service targets for each of the categories A, B and C? If, for example, the overall target is 95% service and the individual target for non-stock items is near zero, then this must mean that the target for A items must be in excess of 95%.
- Suppose an overstock position develops. Is it primarily in the A lines, in the non-stock lines or across the board? It is necessary to know where the bulge is if effective remedial action is to be taken – the remedy for an overstock of A items will differ greatly from that needed to cope with excess stock in the non-stock category.
- While for various reasons it may be best to present one overall service

figure, there may be a need to distinguish between the two causes of non-supply. A customer may not get what he wants either because the company is temporarily out of stock of a stock item or simply because they have quite deliberately decided not to stock it. What is the relative contribution of each of these factors?

To be usable, feedback on stock value and service must be presented in terms of the categories of stock control, in this case non-stock, A, B and C. Unless this is available, then the situation is impossible to manage in any precise manner. Yet many, if not most, stock control systems are poorly served in terms of feedback systems designed to yield information in terms of the categories of control. It's like fire-warning systems. A good system doesn't only tell you that you have a fire – it also lets you know where the fire is.

The measurement of service

The measurement of inventory valuation presents few problems and is not really a matter of debate; the basis of such valuation is readily agreed and accepted. But the measurement of service is much more difficult.

For the most part, service is measured in terms of orders serviced from stock as a percent of total orders received, but this is variously interpreted as number of orders, number of lines, number of items or value of orders. From a strictly theoretical point of view, service should be measured separately for each item, but this is often neither practicable nor necessary. But, as we saw in the previous section, it is necessary to measure service separately for each of the stock control categories – otherwise the situation cannot be fine tuned.

Some situations render the measurement of service relatively easy. A support store for an isolated mine would capture all demands because everything that is ordered must go through the system. This facilitates a very direct measure of service. In other operations, such as those with multiple outlets, for example, the measure of service in terms of ability to satisfy orders is far less reliable. A potential customer may place an order for a scarce item with several branches, even though he only wants one of them. Or a customer may cancel a whole order if he finds negative availability on some of the items. A store whose prime function is to support a workshop servicing heavy plant may find spurious demand arising from a mechanic ordering a kit of likely parts in advance of a job, several of which he does not need and eventually returns to the store.

The point is that service in many operations is difficult to measure reliably. Various attempts have been made to remedy this – notably systematic approaches to defining and capturing 'lost' sales – but few work well in practice. The simple truth of the matter is that in some operations service is easy to measure and in others it is not, and in the latter case the problem is not a system one but is germane to the nature of the business.

Sometimes it is valid to use measures like the number of backorders or the number of items out of stock to give an ongoing indication of service. In both cases, non-stock items should be excluded from the measurement.

When service is discussed, the figures that are bandied about are of the order of 90% or 95%. In practice, this is the type of service offered for the fast-moving bread and butter lines in most operations; but the service in slow-moving lines is much lower – often 60% to 75%. Because most sales come from the fast movers where the service is high, any measure of overall service is also high. Many managers and operating staff do not appreciate just how low is the service they are offering in the slow-moving lines. This point is particularly relevant in terms of setting service objectives for such items; to set them as high as the A targets can get the company into a heap of trouble.

Other measures of feedback

Thus far, we have concentrated attention on feedback of the basic outcomes of the stock control process – investment and service. But there are other things that also need to be monitored. The most obvious is the extent to which the system suggestions are followed. In one company, management felt well pleased with their 'system' because the service was high and the stock was well within budget; but on investigation it was found this was only so because the stock controller totally ignored the system suggestions. In my view, he had good reason to do this – the suggestions were mostly inappropriate – and he seemed to have a feel for what he was doing, which served him in good stead. But management, simply because they had never bothered to look, assumed that he was following the system, and because the results were good, they concluded that the system was functioning well. They got a shock when this man retired.

But the more common difficulty, and one which is particularly pertinent to the concepts of system fine-tuning that we have been discussing, is for stock controllers to overrule the system suggestions wihout due reason. This is extremely common. Yet in such circumstances it is not at all easy to ascertain whether the results – good or bad – are due to the system or to the intervention. So it is highly desirable to have some feedback on the extent to which stock controllers overrule the system and on the reasons for them doing so.

In most cases, it is simple enough to program the computer to report on the number of overrulings compared with the number of acceptances, although this is rarely done. This information, like all the stock feedback data, could also be presented separately for each stock controller, which enables a comparison of performance between the controllers. But while such an analysis shows the number of times the system is not followed, it does not give any indication of the validity of the overrulings. The best way to judge the appropriateness or otherwise of the stock controller's intervention is, from time to time, to go through a suggested reorder report with him after he has processed it, discussing with him the reasons behind any of the changes he has made. It is not always easy to arrive at definite conclusions about each case, but such an analysis leads to an overall impression that is reasonably trustworthy.

It is possible to arrive at a set of rules that, for most items, most of the time, give suggestions that can be followed in 85% to 95% of cases. If this is happening, then you have a stock control system in operation.

Using feedback

The basic idea is to use feedback on system operation to refine the rules and numbers used for control. To do this, we need to begin with initial settings of these rules and numbers. With these in place, the stock system goes into action.

Once it is operational, we must ensure that overrulings are minimized, except in cases of specific market knowledge. Then we know that the results are due to the system, and, if we measure the stock investment and service from each of the categories we are using, we can constantly refine the rules and numbers, and so control stock better.

It is also beneficial to present feedback on investment and service, as well as on the number of overrulings by stock controller and by stock control category. This not only enables a comparison of performance between controllers, but also serves a very important role in the process of learning what works and what does not. It also allows the manager to experiment with different policies with different controllers – a change in policy, for example, can be tried out on a small section of inventory, rather than being implemented across the board.

The role of managers and stock controllers in the stock control process

Communication

In one company, I talked with a manager who had been responsible for the design of the reordering module of the stock control system. With a trace of bitterness, he told me how the stock controllers had initially used his system to great effect but more recently had drifted away from it. If only they had continued to use it, he said, they would have avoided the present overstock position – which was the prime reason for my presence. When I talked to the stock controllers, I heard a completely different story. In their view, the system was incomprehensible – a word they used to describe it to me but not, I suspect, to the manager – and many of the suggestions did not make sense. They showed me a few and I was forced to agree with some of their reservations – I was unable to check the mathematics completely because the written description of the method used by the computer, which was given to me, was not up to date.

This was scarcely a meeting of minds. Rather, there was a collision of diametrically opposed opinions on whether the system 'worked' or not. There were elements of truth on both sides, but there were no communication structures in place to explore the issue and develop a workable way to move forward.

Many marriages come to a stage where the partners are unable to communicate. They desperately want to, and indeed frequently try, but the emotional baggage they have accumulated gets in the way and discussions degenerate into arguments. The Marriage Guidance practice in such situations is to attempt to establish a contract or agreement between the parties on key points of contention – for starters, the husband agrees to go drinking with the boys only twice a week instead of five times, while the wife agrees to have the house and herself tidy by the time he arrives home after work. The basic idea is to gradually redirect the energy spent on emotion to constructively rebuild the relationship.

There is, of course, no intention here of implying that the relationship between the manager and the stock controller possesses the emotional overtones of a failing marriage. But there is often an analogous problem of communication in situations like the example quoted. Both managers and stock controllers want the stock control process to work and both parties have contributions to make, but the difficulty is to use this expertise, knowledge and energy to gradually but consistently improve the operation. They reach a stalemate and can progress no further. It is common in my experience to revisit a company after an absence of a year or two or three and to find the same people doing the same things in the same way as before. There is undoubtedly a desire to improve but an inability to translate this to action.

What is needed in such cases is a means of structuring the situation so that communication is facilitated. In practical terms, this means clearly defining rules and procedures and agreeing to work within their framework while at the same time maintaining flexibility; if there is a problem, the option is not to abandon the structure but to work within it to resolve the difficulty.

Clarity and consensus

In setting up the stock control system, the initial categorizations, rules and parameter settings should be worked out jointly by the stock controllers and the managers. In this area, it is very easy to bulldoze a system through – especially if it contains unfamiliar mathematical algorithms – simply because one person is committed to it. Stock controllers may be reticent about expressing their true opinions in the presence of their superiors, with the effect that a lack of verbal objection is taken as agreement. It is not uncommon for the stock controllers to have little or no say in the stock system design – and this doesn't only apply to packages.

If a stock controller doesn't like a system, he is more likely to overrule it than try to change it. This is especially so if the change requires programming. If he does discuss it with management and the reaction is not positive, then he is unlikely to bring the matter up again. But this climate is not appropriate if the objective is to refine the system by learning its strengths and weaknesses. User reactions are valid and need to be discussed in a non-threatening climate, and a course of action agreed – either to change the parameters or for the user to follow the system despite his reservations. The action may prove to be wrong but it is possible then to learn from the experience.

In this area, we could learn much from the Japanese management style which goes to great pains to ensure that the issues are clear and that a consensus of all affected parties is achieved before a decision is made. This process takes time and the initial reaction of many western managers exposed to this for the first time is that of endless discussions which seem to be going nowhere. But it mostly results in a course of action that has been thoroughly explored in advance and with the expectation that everyone will go along with it fully, once it is implemented. And this is what happens.

In the West, we tend not to get down to details in our discussions of planned actions, especially relating to stock control. As a result, we often operate using vague rules which only guide the ordering process in a fuzzy way. This approach doesn't work in the fast-food business where survival has been shown to depend heavily on clearly defining and ensuring compliance with every aspect of restaurant operation; nor does it work in stock control.

This nebulous approach often comes to the forefront in meetings on stock-related issues. I have attended many meetings like the one that was called by a health food importer which had two items on the agenda. The first of these was why they were out of stock of their single best-selling line, a coffee substitute. The second topic was the managing director's concern with the fact that overall stocks were 10% above budget. The meeting went on for hours. Much of the time was spent in people giving the reasons that they thought caused the present position, and arguments rose and subsided as the blame was passed around. The meeting did not, as far as I could tell, take any further decisions than to fly in some coffee substitute and to cancel some big orders. But the first had already been put in place before the meeting, and it took only a brief look at the stock balances and sales patterns to ascertain that the second set of decisions was dubious at best. The meeting wasted a lot of time and achieved nothing.

In another company with overstock, the manager and the stock controller met for 5 minutes and decided to react by reducing the safety stock settings for B and C lines by a week. They weren't completely sure of what the effect would be, but some simple calculations indicated that this should bring them back on budget in a couple of months. They were able to decide on actions directly related to the problem because they had clearly defined structures in place and they had agreed to exercise control through these structures. It's so much easier and less fraught with emotion to do it this way.

It is also more logical. It doesn't take much thought to appreciate that common reactions to overstock are often illogical. In one company where the stock controller was under pressure to reduce stock, he adopted a policy of ordering only 75% of any suggested order; this is quite simply inconsistent. In another relatively small operation, the managing director became so exasperated with what he saw as his stock controller's inability to control the stock that he insisted on personally reviewing any order placed with a monetary value in excess of £300.00; his intervention was erratic at best because he knew little detail of most of the products, and it wasted a lot of his time. One order was delayed 2 weeks while waiting for him to return from a vacation.

Supplier relations

Recent developments

In recent years, the impact of the manufacturing revolution, known as Just-in-Time, and the advent of enhanced communications facilities has meant that the supplied–supplier relationship has changed drastically. With high-volume, mostly repetitive, components in automotive assembly, it is now commonplace for suppliers to deliver several times a day; the assembly plant holds only a few hours supply of such items at any time.

Such developments are often seen as perhaps less relevant to warehouse operations with thousands of items and long lead times. But like most such innovative concepts, the supplier relation is something that needs to be worked out for the individual company, and in doing this it is possible to learn from what others have been able to achieve. Perfection comes not from slavish imitation of well-publicized models, but from learning the extent of the possible from the practice of others and applying the underlying applicable concepts to our own operations.

Front end lead time extensions

In any stock operation, it is important to communicate our needs to the supplier as quickly as possible after they have arisen. Many of us are not as quick on our feet as we might be in this matter, and it is not uncommon for the delay to be at least partly of our own making. Of course, it helps if we have state-of-the-art systems.

Airline spares have been ordered for years from prime suppliers using on-line links, and this is now spreading to other types of operations. A heavy plant maintenance operation in the UK has a system that automatically calculates replenishment requirements at the end of each day, and these are transmitted overnight and appear as picking lists the following morning in the supplying plant in Amsterdam. They are picked and sent before midday and arrive the following day.

This contrasts with another spares operation which consists of a central warehouse and in excess of 50 branches. A central computer processes all transactions and performs accounting functions in batch mode; paperwork is sent in once weekly from each branch. A suggested reorder report is produced each week and sent to the branch, and they review it and place their orders on the central warehouse. At first sight, this seems OK. But a closer inspection of the time this total internal supply takes illustrates the problem. Consider this scenario where the days are working days:

- Day 1: Close off weekly accounts at end of day. Paperwork posted to central computer.
- Day 4: Paperwork punched up. Overnight a suggested reorder report is produced. Posted to branch the following morning.

- Day 12: Closing day for receipt – by post – of revised suggested reorder report from the branch.
- Day 13: Changes punched up. Overnight picking lists are produced and the central warehouse picks, packs and dispatches within 3 days.
- Day 19: Goods arrive at branch and are unpacked and entered into stock.

This means that a demand occurring in the week before close-off (Day 1) is replenished by the arrival of goods on Day 19 – a total lead time of between 4 and 5 weeks. In the previous example, this process took less than 3 days.

In practice, no one at the branches liked the system because it was so out of date. They spent inordinate amounts of time updating the suggested reorder report from their manual records of the demands that had occurred since the close-off a week earlier. The main reason for changing – or adding to – suggested orders was the fact that the item had moved in the week since the report was produced; it had nothing to do with market knowledge.

While there were considerable differences between the two examples, in terms of number of branches and systems in place, the elapsed time in the second example was excessive. It could have been – and subsequently was – reduced from 4–5 weeks to 1.5–2 weeks, simply by eliminating the review by the branch, using overnight couriers and smartening up the warehouse response – that is, there was no change in the basic systems. The revised timing was as follows:

- Day 1: Close off weekly accounts at end of day. Paperwork sent by overnight courier to central computer.
- Day 2: Paperwork punched up. Overnight a picking list is produced. This is picked, packed and dispatched before the end of day 3. Overnight transport used.
- Day 4: Goods arrive at branch and are unpacked and entered into stock.

The movement towards automatic replenishment – where the opportunity to review the order before it is placed on the supplier is eliminated – is growing for the simple reason that the review often slows down the process quite considerably. Clearly, this option must be based on a good grounding in terms of the rules in place that suggest the orders. If many suggested orders are overridden for reasons that have nothing to do with market knowledge, then such a system would be a poor basis on which to mount automatic ordering.

But the point remains that, in many operations, there are internal delays in either recognizing a need to replenish or in communicating it to the supplier. This problem is compounded by the wide lack of recognition on the part of management that there is a problem. In the second case mentioned earlier, everyone – branch personnel, central warehouse, systems and management – argued that the original timings reflected nothing more than the minimum amount of time the various links in the chain of supply needed to do their jobs; there was little recognition of the cumulative effect, and in fact the elapsed timings came as a surprise to many, not because the elements were unknown to them but simply because they had never thought about it from that point of view.

Supply arrangements

With a large product range the normal method of replenishment is to place an order on the supplier when the order point is reached. Sometimes the supplier is provided with a schedule which specifies not only immediate offtake requirements but also expected future needs. For many if not most items in the range these are probably the most appropriate approaches to the communication of supply need. And, as we saw in the previous section, the speed of this response is a key determinant in the balance that is maintained between investment and service.

But there is also a place for developing more detailed arrangements for the supply of some items, especially high volume lines or lines which have difficulties of supply. These are the types of items that are well worth discussing in some detail with the supplier to see whether there are avenues that can be explored to mutual benefit. In such discussions there is a need for frankness. If there is an agreement then it should be interpreted literally and there should be a high level of dependence on it. The two modes of operation described below illustrate this point.

A supplier makes delivery promises that in many cases he fails to meet, but he still delivers most – 75% – of his orders on time. Also, his customers often change their minds within the lead time, pulling some items forward and pushing back or even cancelling others. So the supplier holds buffers to cover himself against the customer changing his mind, while the customer routinely asks for things a week early to cover himself against what he sees as a supply of doubtful reliability.

But suppose both sides meant what they said. The customer's delivery date is the day his customer needs it or the day in which it will be used in production – it will be a catastrophe if this delivery is missed as there is no back-up. The supplier knows this – and also knows that this is a firm requirement that will not be changed up or down in the time before delivery – so that meeting his delivery is not just a priority but an absolute priority. In such circumstances, where the conditions of supply are crystal clear and any failure to abide by these conditions precipitates a crisis, there is a high likelihood of compliance. This is of course the way of JIT, but the concept is applicable to warehouse operations. Buffers of time or material that mean survival in the event of missed deliveries are put in place to guard against such events. But their effect is to de-emphasize the importance of compliance so that missing a delivery is not seen as a serious matter, which in turn means that due dates don't mean much.

There is a universal truth here that can also be illustrated by considering an example from another aspect of supply. A large Australian company imported electrical goods from Italy. The documentation took much time to prepare and had to be accurate, or there were hold-ups. One of the Australian company's staff was fluent in Italian and he gradually became involved in sorting out these difficulties, a task that he did superbly and quickly. As time passed, a curious pattern emerged. The quality of presentation of the original documentation degenerated and the time spent by this man in sorting it out increased to the extent that he eventually headed up a department of two people devoted solely to this task. Because there was now no real need to get the documentation right first time – it was no longer a cause of delay – the accuracy was not seen as a prime requirement by those concerned in document preparation.

In the same manner, a way – possibly the only way – to focus attention on ensuring compliance with the arrangements of supply, on the part of both the supplier and the supplied, is to structure the situation so that failure is intolerable. If, for our key products, we wish to explore special conditions of supply, then this could well be the guiding principle. The emphasis should be to build trust rather than buffers.

There is enormous variety in the nature of the supply conditions that best suit individual key products. The customer's ordering pattern may be tied in with the supplier's pattern, so that any needs are reflected quickly to further sources of supply. A Kanban approach might be used – where the supplier knows in advance the size of any order and only the frequency varies. If the supplier is a manufacturer, and especially if the product is made specifically for the customer, then the setting of batch sizes and the extent of commitment to any remaining items from a batch could well be matters for discussion and formal agreement. Or, if the supply is difficult or long, then there may well be a case for holding buffers of one sort or another – but preferably only in one agreed place. Because of the inclination to treat all items in much the same way, there is a tendency not to exploit the supply conditions to the best benefit of both the supplier and the supplied.

In conclusion

There is more managing an operating stock situation than the three topics discussed in this chapter. But if a manager is guided by adequate feedback, is able to harness his own and his stock controllers' energies so that progress is ongoing, and works at exploiting the supply interface to the mutual benefit of the supplier and himself, then he is doing more than most.

CONCLUDING REMARKS

Effective stock management means achieving and maintaining the best possible balance between investment in stock and service to customers. This cannot be achieved simply by using global measures of performance on these two dimensions to exert pressure on the stock controllers to rectify problems that have already occurred.

Stock management is a matter of detail. For starters, the manager needs to have a solid appreciation of what is involved in the process and of the difficulties that tend to arise in its performance. He also needs to ensure that his plans and actions are coordinated with, and give due regard to, the thinking of the stock controllers.

Stock management should be systematic. This means that the ordering process should be guided by clearly defined rules which are overridden if, and only if, there exists some specific quantifiable market knowledge relating to the product. These rules should be simple rather than complex, and understood and accepted by both the managers and the people who do the ordering. This is the system design phase and it is up to the manager to see that it happens.

But the prime test of management comes when a stock system becomes operational. Does the process remain systematic or does it degenerate into a vague mess of subjective input in which the order depends on who does the ordering? As time passes, are the rules fine tuned and modified in the light of gained experience, or do they gradually lose their credibility by being neglected and not updated? Is the management process one of sudden occasional reactions to crises that should never have been allowed to develop, or is it a hands-on continuous learning of what works and what does not?

In this management process, the necessity of feedback – by control category and possibly by buyer – is obvious. You cannot go where you want unless you can see where you are going. Because there is so little emphasis on feedback tailored to stock management in most companies, it is no wonder that this management goes astray.

The emphasis of Just-in-Time is summed up in concepts like 'habit of improvement' and 'progressive elimination of wasteful practices'. These ideas hold considerable validity for inventory control; there is much waste and enormous potential to improve. Whether these potential savings are realized or not depends on the quality of stock management.

APPENDIX 1

COMMON ALGORITHMS USED FOR FORECASTING AND INVENTORY CONTROL

Overview

This appendix sets out the more basic algorithms used in forecasting and inventory control. Despite the fact that they are very much in the lower reaches of model sophistication, the algorithms covered are the ones that tend to be used in practice, where any is used at all.

The discussion centres on the four parameters of any inventory control system:

1. Forecast.
2. Safety stock.
3. Order quantity.
4. Lead time.

Forecasting

Averaging and exponential smoothing

The most common methods of forecasting used in stock control systems are simple averaging and exponential smoothing. Both attempt to estimate future demand from past historical demand.

If D_t is the actual demand in period t and F_t is the forecast for period t, then the simple average is calculated as follows:

$$F_{t+j} = \left(\sum_{i=T-N+1}^{T} D_i \right) / N$$

where T is the most recent actual period, N is the number of periods on which the average is based and $j = 1, 2, 3, \ldots$.

Using exponential smoothing, the new forecast is calculated by updating the previous forecast by a percentage of the error in that forecast:

$$F_{T+j} = F_T + \alpha(D_T - F_T)$$

or:

$$F_{T+j} = \alpha D_T + (1 - \alpha)F_T$$

where α is the smoothing factor $(0 \leqslant \alpha \leqslant 1)$.

But F_T could also be expressed in terms of the previous forecast and demand:

$$F_T = \alpha D_{T-1} + (1 - \alpha)F_{T-1}$$

from which:

$$F_{T+j} = \alpha D_T + \alpha(1 - \alpha)D_{T-1} + (1 - \alpha)^2 F_{T-1}$$

Proceeding in this manner, it is possible to express the exponentially smoothed forecast in terms of a weighted sum of past demands:

$$F_{T+j} = \alpha D_T + \alpha(1 - \alpha)D_{T-1} + \alpha(1 - \alpha)^2 D_{T-2} + \alpha(1 - \alpha)^3 D_{T-3} + \ldots$$

Hence, the term exponential.

In both of the above cases:

$$F_{T+j} = F_{T+1} \qquad \text{for } j = 1, 2, 3, \ldots$$

as neither of the methods takes account of either seasonality or trend.

There are two variations on the above that deserve mention.

A weighted average is similar to a simple average except that different weights are given to past demands. The normal pattern is for more recent months to carry greater weights. The weighted average is calculated as follows:

$$F_{T+j} = \left(\sum_{i=T-N+1}^{T} W_i D_i \right) \Bigg/ \left(\sum_{i=T-N+1}^{T} W_i \right)$$

where W_i are the weights.

Adaptive smoothing is a variation on exponential smoothing in which the weighting factor, α, depends on the pattern of variation of past demand. One approach to this is to set α equal to the tracking signal (TS) where:

$$TS = \left| \frac{MSD_{T+1}}{MAD_{T+1}} \right|$$

and the mean signed deviation (MSD) and mean absolute deviation (MAD) may both be calculated using exponential smoothing:

$$MSD_{T+1} = (1 - \beta)MSD_T + \beta(D_T - F_T)$$

$$MAD_{T+1} = (1 - \gamma)MAD_T + \gamma|D_T - F_T|$$

Accounting for trend and seasonality using simple averaging

There are multiple approaches to extending the concept of an average to take account of trend or seasonality or both. Let us consider the concept of trend first.

The technique favoured here is to use the least squares approach to deriving the line of best fit. If a linear model is used, then the equation of the line of best fit is:

$$F_i = a + bi$$

where i covers both past and future periods.

The values of a and b are readily derived:

$$b = \frac{N\left(\sum_{i=T-N+1}^{T} iD_i\right) - \left(\sum_{i=T-N+1}^{T} i\right)\left(\sum_{i=T-N+1}^{T} D_i\right)}{N\left(\sum_{i=T-N+1}^{T} i^2\right) - \left(\sum_{i=T-N+1}^{T} i\right)^2}$$

$$a = \frac{\left(\sum_{i=T-N+1}^{T} D_i\right) - b\left(\sum_{i=T-N+1}^{T} i\right)}{N}$$

where N is the number of periods used in the calculation of the regression line and T is the most recent actual period of demand. (These equations are derived in most elementary texts on statistics and forecasting and will not be derived here.)

The forecast for period $T + j$ is then calculated:

$$F_{T+j} = a + b(T + j) \qquad j = 1, 2, \ldots$$

Non-linear models may readily be used to estimate the trend line but there is little suggestion that they are of relevance to forecasting for inventory control purposes.

Averaging may also be extended to take account of seasonality. For simplicity, we will assume that periods are months and the season is a year.

The most common approach used to estimate seasonality is through the application of seasonal indices. If several years demand history are available for a product, then the seasonal index for each month may be calculated thus:

$$S_l = \frac{\left(\sum_{k=1}^{M} D_{lk} \right) \times 12}{\left(\sum_{k=1}^{M} \sum_{l=1}^{12} D_{lk} \right)}$$

where D_{lk} is the demand in month l of year k and S_l is the seasonal index for month l.

Clearly:

$$\sum_{l=1}^{12} S_l = 12$$

The forecast for any period is then:

$$F'_{T+j} = F_{T+j} \times S_{T+j} \qquad j = 1, 2, 3, \ldots$$

where F_{T+j} is the average calculated earlier, F'_{T+j} is the new forecast which takes account of seasonality and S_{T+j} is the seasonal index for month $(T + j)$.

This is a simple approach to calculating seasonal indices from past history and is inappropriate if there is an underlying trend. In such cases, percentages of a years' demand may be used, instead of actual demands in the above calculations of indices, but this approach also has deficiencies.

But the prime difficulty in the calculation of seasonal indices is the often high-random component in patterns of demand at the item level.

This often nullifies any attempt, no matter how sophisticated, to calculate such indices in this manner.

If both seasonality and trend are present, then the usual approach is to proceed as follows:

1. Calculate the seasonal indices.
2. Deseasonalize the time series by dividing each month's demand by the relevant seasonal index.
3. Extrapolate the resulting time series by using the least squares technique.
4. Multiply these extrapolated monthly figures by the seasonal indices to yield the forecasts for each month into the future.

There are a multitude of variations on the basic theme in the various textbooks on forecasting and, in practical terms, there are many computer packages that forecast using trend and seasonality to a greater or lesser degree of sophistication.

This discussion has not attempted to raise any of the real theoretical issues relating to time series analysis, and the algorithms presented are of a very simplistic nature and could be criticized as inadequate on numerous counts. But they represent a crude but reasonably effective manner of taking account of trend and seasonality for inventory control purposes without undue mathematical complication.

Accounting for trend and seasonality with exponential smoothing

The general model here assumes that the underlying average demand, the seasonal factor and the trend factor are all determined using simple exponential smoothing. As with the previous approach, trend is additive while seasonality is multiplicative.

The forecast for period $T + j$ is:

$$F_{T+j} = (A_T + jT_T)S_{T+j-12}$$

where A_T is the exponentially smoothed average based on periods up to and including period T, T_T is the trend factor and S_{T+j-12} is the seasonal factor. All three of these components are calculated using exponential smoothing, but modified because of the intrusion of the other components.

The average component, A_T, is calculated by:

$$A_T = \alpha \frac{D_T}{S_{T-12}} + (1 - \alpha)(A_{T-1} + T_{T-1})$$

which discounts the seasonal component but takes account of the increase due to trend.

The seasonal factor is:

$$S_T = \beta \frac{D_T}{A_T} + (1 - \beta) S_{T-12}$$

which attempts to isolate the seasonal factor by dividing the latest demand by the 'average'.

Finally, the trend factor is:

$$T_T = \gamma(A_T - A_{T-1}) + (1 - \gamma)T_{T-1}$$

which updates the trend with the latest change in the underlying average.

As each new demand becomes known, A_T is first calculated, then S_T and T_T; these are then used to calculate F_{T+j} for periods $j = 1, 2, 3, \ldots$. Clearly, the approach can be modified to ignore either seasonality (by setting $S_i = 1$, all i) or trend (by setting $T_i = 0$, all i).

Further forecasting algorithms

The algorithms just presented represent the bare bones of some of the simpler approaches to generating forecasts. Much has been omitted – tests of significance, for example – that are germane to forecasting at any level. But given the high-random component in most patterns of demand at the item level, the algorithms set out represent an adequate if crude approach to the forecasting of item demand for the purposes of inventory control.

There is an abundant literature on forecasting and time series analysis, and the reader is advised to consult one or other of these more specialized books if he wishes to explore the topic in greater detail.

Setting safety stocks

Safety stocks may be set in terms of units of weeks of stock. Such approaches do not take account of variation in demand nor are they able to be directly related to theoretical constructs like the desired service level.

If safety stock is set in terms of standard deviation (SD) or mean absolute deviation (MAD), then this implies higher safety stocks for more variable patterns of demand. The MAD has little theoretical justification, being introduced in the first instance to speed up calculations and/or obviate the necessity of finding square roots on a generation of computers on which calculations were relatively slow and square root algorithms, in at least some cases, not available.

The formulae for SD and MAD are:

$$\mathrm{SD}_{T+1} = \sqrt{\sum_{i=T-N+1}^{T} (D_i - A_{T+1})^2 \Big/ (N-1)}$$

where:

$$A_{T+1} = \left(\sum_{i=T-N+1}^{T} D_i \right) \Big/ N \quad \text{and}$$

$$\mathrm{MAD}_{T+1} = \sum_{i=T-N+1}^{T} |D_i - A_{T+1}| \Big/ N$$

There is a tendency on some systems that use exponential smoothing to calculate the MAD using the following technique:

$$\mathrm{MAD}_{T+1} = (1 - \alpha)\mathrm{MAD}_T + \alpha|D_T - A_T|$$

In the following, we will use the SD as the measure of variation. In practical, if not theoretical terms, the MAD might be used instead.

The simplest method of setting safety stocks is in terms of 'so many' SD,:

$$\mathrm{SS} = X \times \mathrm{SD}$$

This approach takes no account of lead time and yet it is clear that higher safety stocks need to be set for higher lead times to achieve the same level of service.

The approach that tends to be used commonly here is to assume that the SD of demand in a period of time varies as the square root of the length of that period. Applied to lead time, measured in months, and SD, calculated for a month, then the SD during the lead time would be calculated as:

$$\mathrm{SD}_{\mathrm{LT}} = \mathrm{SD} \times \sqrt{\mathrm{LT}}$$

where LT is the lead time in months.

There is no theoretical justification for this precise formulation, and it is still a matter of some debate, but it is clearly preferable to the extreme alternatives, as in the following:

$$\mathrm{SD}_{\mathrm{LT}} = \mathrm{SD}$$

$$SD_{LT} = SD \times LT.$$

Thus, the safety stock may be also calculated as:

$$SS = X \times SD_{LT}$$

to take account of lead time.

But the most common use of SD in the setting of safety stocks is to use a model based on a desired service level (DSL). There are many of these derived for various definitions of service and cost structures. We will consider only one here which is commonly used.

This is based on the distribution of the forecast errors rather than on the distribution of demand. Moreover, the model assumes that these are calculated over a lead time (assumed fixed) rather than over a period like a month.

The safety stock in the order–point formula is determined as 'k' standard deviations. Clearly the setting of k determines the service. The relation is established using the partial expectation function of whatever demand distribution is assumed. This is a measure of the average backorder quantity and is defined by:

$$f(k) = \int_k^\infty (x - k)\, p(x)dx$$

where $p(x)$ is the distribution. Tables of the partial expectation function are available for various distributions.

The process for determining safety stock is simple enough. First, a value of f(k) is calculated:

$$f(k) = \frac{Q}{\sigma}\ (I - P)$$

where Q is the average reorder quantity, σ is the standard deviation of the forecast errors during the lead time, and P is the desired service level.

Using the table of the partial expectation function of whatever distribution is assumed, the corresponding value of k can be found. The safety stock appropriate to the desired service level, P, is then $k\sigma$.

Economic order quantity

There are many versions of the economic order quantity (EOQ) derivation. We will pursue only the simplest and the oldest because it is the only one used widely in practice.

The basic assumptions are that there are two costs associated with ordering, the cost of placing an order and the cost of holding stock.

Over a period of a year, the total cost of placing orders, at a cost of £A per order, would be:

AT/Q

where T is the total demand for a year and Q is the order quantity.

The holding cost depends on the average inventory. Assuming the order arrives on average as safety stock, S, is reached, then the average stock is:

$S + Q/2$

The carrying cost is this multiplied by unit cost C and the percentage carrying cost P, which is a single figure that reflects costs like insurance, wages, interest, and so on.

Thus, the holding cost is:

$(S + Q/2) \, CP/100$

The total cost is:

$$(S + Q/2) \, C \, P/100 + \frac{AT}{Q}$$

Differentiating with respect to Q, we have:

$$\frac{CP}{2 \times 100} - \frac{AT}{Q^2}$$

It is readily shown that this is a minimum. The value of Q for this minimum is:

$$Q = \sqrt{\frac{200 \, AT}{CP}}$$

This is the EOQ value.

Lead time smoothing

This is simply exponential smoothing applied to updating lead time:

$$LT_N = (1 - \alpha)LT_0 + \alpha LLT$$

where LT_N is the new calculated lead time, LT_0 is the previous calculated lead time and LLT is the latest actual lead time.

In conclusion

The formulations presented here are quite trivial from the mathematical point of view, but they cover the prime algorithms used in most practical inventory operations.

There is in fact an enormous variety of alternative more complex and more comprehensive algorithms in the operations research literature, but these are little used in practice. The prime reason for this would seem to be that the random component is so high in most patterns of demand at the item level that the advantages of discovering and using underlying patterns of demand or of optimizing in any real sense are difficult to realize in practice.

APPENDIX 2:
CALCULATION OF TOP-UP LEVEL FOR MONTHLY ORDERING

Consider a situation where orders are placed once monthly, at the same time each month, and the lead time is 6 weeks. For the present, we will assume there is no allowance for safety stock. Assuming 4 weeks to the month and that orders are placed and received on the first day of any week, we have the pattern of orders and receipts shown in Figure 9.

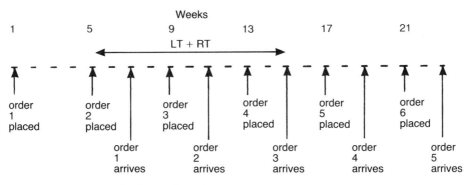

Figure 9 *Pattern of orders and receipts.*

Consider 'order 2' which is placed in week 5 and arrives in week 11. This receipt, plus any stock on hand, needs to last until the arrival of 'order 3' in week 15. So 'order 1' must last until the arrival of the next order – a total of 10 weeks. This is the level up to which stock needs to be raised at the placing of 'order 2' – that is the 'top-up level' is 10 weeks. This is the sum of the lead time (6 weeks) and the review time (4 weeks). To cover against lead time extension or high demands during the 10 weeks, some safety stock would also be carried.

APPENDIX 3

CALCULATION OF SAFETY STOCK AS A PERCENTAGE OF TOTAL STOCK

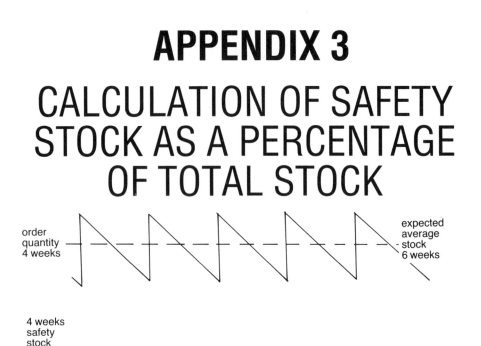

Figure 10 represents a typical stock situation. It only illustrates the stock on hand – the stock on order is excluded. If the lead time is 12 weeks, and the MIN is set at 16 weeks and the MAX at 20 weeks, then the safety stock is 4 weeks. Moreover, the average order quantity is also 4 weeks. The stock fluctuates between the safety stock (4 weeks) and the safety stock + Order Quantity (4 + 4 = 8 weeks): the expected average stockholding is half-way between these extremes:

Expected average stock = safety stock + 1/2 order quantity
= 4 + (1/2 × 4) = 6 weeks

Safety stock as fraction of total stock = safety stock/expected average stock
= 4/6   or 67%

Index

ABC classification, 31, 64, 66
ABC/FMS categorization, 64-6, 68-70
automatic calculation, 14
automatic update of ordering parameters, 16
averaging, 40, 68, 89-93

backorders, 77
Brown, R G, 27
budgetary control, 68

categorization process, 61-70
    determining rules and parameters for, 67-9
    stock and non-stock items, 61-3
communication issues, 78–80
computer systems, 13, 15, 35, 37, 55, 60

demand fluctuations, 71-2
demand patterns, 20
Deming Cycle, 75
design of system, 57-72
    stock/non-stock division, 65-6
desired service level (DSL), 96

economic order quantity (EOQ), 31, 58, 59, 69
    extent of usage of, 53
    formulation of, 52-3
    mathematical algorithms, 97-8
equations, 25-6, 68
experience,
    argument for, 19-20
    importance of, 19-21
    role of, 20
    systematic approach combined with, 22
    vs system, 21-2
exponential smoothing, 43-6, 54, 58, 59, 68, 89-91, 93-4
    lead time, 97-8

feedback,
    absolute necessity of, 74-6
    for fine tuning, 73-8
    importance of, 73-4
    measures of, 77-8
    utilization of, 78
FMS classification, 65, 66
    see also ABC/FMS
forecasting, 21, 32, 40, 52, 68, 71
    alternative strategies of, 55
    and averaging, 40
    mathematical algorithms, 89-94
    methods of, 41-3
    month-by-month, 47-8
    summary of methods, 48

inventory control,
    categorization, 61-6

classifications, 63-4
decisions of, 26-7
prime act of, 9
rules for see rules
sophistication in, 54-6
systematic approach, 19
    combined with experience, 22
    definition and use of term, 9-11
    practical examples, 13-18
    vs experience, 21-2
test case, 10
investment and service levels, 36-7

Just-in-Time (JIT), 27, 35-6, 81, 85

Kanban, 27, 36, 59

lead time, 19, 20, 30, 32, 40, 52-4, 71, 82-4, 95
    determination of, 53-4
    exponential smoothing, 97-8

management of system, 73-86
managers, role of, 78-81
market knowledge, 19, 20
    in ordering strategy, 22
    problems of, 20-1
materials requirements planning (MRP)
    systems, 27, 34-5
    as solution to manufacturing, 35
    dependent and independent demand, 34
    logic of, 34-5
mathematical algorithms, 16, 27, 54-5, 58-9, 89-94
    economic order quantity (EOQ), 96-7
    forecasting, 89-94
    safety stocks, 94-7
mathematical models, 30
mathematics, 16, 39-56
    and lead time determination, 53-4
MAX/MIN systems, 13-14, 27-8, 33, 49, 53, 58
mean absolute deviation (MAD), 30, 50-52, 95
message displays, 72
MTBF (mean time between failures), 68

one-level systems, 27, 32-4
order point,
    concept of, 27-8
    determination of, 28-9
    equation for, 68
order point/order quantity (OP/OQ) systems,
    27-8, 31, 52, 59-60, 67, 68, 70
order quantity,
    approaches to setting, 31, 32, 40
    calculation of, 69

order-up-to level (OUTL), 33-4
ordering process,
    levels, 60, 67
    market knowledge in strategy of, 22
    patterns, 11
    rules for, 11-12
    subjective element in, 11-12
ordering windows, concept of, 32-3
out-of-stock situation, 17, 77
overrulings, 23, 77-8
overstocking, 17, 68, 81

parameters, 32, 39-40, 52, 53
    calculation of, 54
    clarity and consensus on, 80
    determination of, 55-6
        for ordering categories, 67-9
Pareto class, 63
periodic review system, 33
Plossl, George, 48

reorder point/reorder quantity approach, 15
reorder report, suggestions in, 23
rules, 10, 12, 25
    categories of, 57-8
    clarity and consensus on, 80
    determination for ordering categories, 67-9
    non-existent or poorly defined, 57
    specification of, 60-1
    stock/non-stock categorization, 62-3

safety stocks, 32, 40, 48-52, 58, 59
    approaches to setting, 49-50
    as percentage of total stock, 100
    calculation of, 49
    control exercised using MAX/MIN
        approach, 49

critical role of settings for, 30-31
    mathematical algorithms, 94-7
    role of, 48-9
    service level approach, 50-52
    setting, 69, 81
        guidelines for, 30
seasonal factor, 93-4
seasonal index, 47, 92
seasonality, 47-8, 68, 71, 91-4
    accounting for, 47
service levels, 11, 17, 18, 23-4, 36-7, 96
    and safety stocks, 50-52
    measurement of, 76-7
spreadsheets, 69
standard deviation (SD), 30, 50-52, 95
stock control, overview, 9
stock controllers, role of, 78-80
stock levels, 11, 17, 18
stock parameter calculations, 16
stock reduction, 17, 18
subjective element in ordering process, 11-12
suggested reorder report, 23
suppliers, 63
    relations with, 81-6
supply arrangements, 84-6
systems
    availability, 25-38
    variety of, 25-7

time series analysis, 40-41
top-up level, 33-4
    calculation of, 99
trend, 71, 91-4
    concept of, 46-7, 68
trend factor, 93-4
two-level systems, 27-32